CLIMATE CHANGE AND ITS CAUSES, EFFECTS AND PREDICTION

IMPLICATIONS OF CLIMATE CHANGE IN MEXICO, THE CARIBBEAN AND CENTRAL AMERICA

Climate Change and its Causes, Effects and Prediction

Additional books in this series can be found on Nova's website under the Series tab.

Additional E-books in this series can be found on Nova's website under the E-book tab.

CLIMATE CHANGE AND ITS CAUSES, EFFECTS AND PREDICTION

IMPLICATIONS OF CLIMATE CHANGE IN MEXICO, THE CARIBBEAN AND CENTRAL AMERICA

SAMUEL B. DUNCAN
EDITOR

Nova Science Publishers, Inc.
New York

For permission to use material from this book please contact us:
Telephone 631-231-7269; Fax 631-231-8175
Web Site: http://www.novapublishers.com

Additional color graphics may be available in the e-book version of this book.

LIBRARY OF CONGRESS CATALOGING-IN-PUBLICATION DATA

Implications of climate change in Mexico, the Caribbean, and Central America
/ editor, Samuel B. Duncan.
p. cm.
Includes index.
ISBN 978-1-61122-849-6 (hardcover)
1. Climatic changes--Mexico. 2. Climatic changes--Caribbean Area. 3.
Climatic changes--Central America. 4. Mexico--Climate. 5. Caribbean
Area--Climate. 6. Central America--Climate. I. Duncan, Samuel B.
QC903.2.M59I67 2010
551.6972--dc22

2010043899

Published by Nova Science Publishers, Inc. † New York

CONTENTS

PREFACE

Mexico, the countries of the Caribbean and Central America are at risk from the impacts of climate change in the next 20 years because they will be exposed to a greater range of climate changes and have a relatively weak adaptive capacity when compared to the world at large. Within the region, climate change is evident in increased temperatures, changes in precipitation, and sea level rise, and perhaps in weather variability and natural disaster events. Countries discussed in this book include Belize, Cuba, the Dominican Republic, Guatemala, Haiti, Honduras, Mexico, Nicaragua, Panama, and Puerto Rico.

Chapter 1- Mexico, the countries of the Caribbean, and Central America examined in this report are at risk from the impacts of climate change in the next 20 years because they will be exposed to a greater range of climate changes and have a relatively weak adaptive capacity when compared to the world at large. Within the region, climate change is evident in increased temperatures, changes in precipitation, and sea level rise—and perhaps in weather variability and natural disaster events. Countries in this report include Belize, Cuba, the Dominican Republic, Guatemala, Haiti, Honduras, Mexico, Nicaragua, and Panama; Puerto Rico is also discussed.

Steady increases within the region in the number of extreme weather events—hurricanes, storms, and droughts—and their effect on infrastructure, public health, loss of human life and agriculture may be attributable to climate change. The countries reviewed do not yet have a full understanding of the potential impacts of future climatic changes and are not prepared to prevent or reduce those impacts.

Chapter 2- The National Intelligence Council-sponsored workshop entitled The Implications of Global Climate Change in Mexico, the Caribbean, and Central America, held on 31 August 2009, brought together a panel of regional experts to consider the probable effects of climate change on Mexico, the Caribbean, and Central America from a social, political, and economic perspective. The panelists concluded that through 2030 climatic changes in the region may aggravate civil unrest and internal conflicts leading to increased migration, and that strong, centralized states, and states with robust civil societies, will likely fare better than others.

Although the region does not contribute to significant global greenhouse gases, it is highly vulnerable to the effects generated by increasing climate variability. Rising temperatures, rising sea levels, increased rainfall in some places, drought in others, and a greater frequency of extreme weather events such as hurricanes, floods, and heat waves are expected from climate change.

In: Implications of Climate Change in Mexico… ISBN: 978-1-61122-849-6
Editors: Samuel B. Duncan

Chapter 1

MEXICO, THE CARIBBEAN, AND CENTRAL AMERICA: THE IMPACT OF CLIMATE CHANGE TO 2030- A COMMISSIONED RESEARCH REPORT

National Intelligence Council

SCOPE NOTE

Following the publication in 2008 of the National Intelligence Assessment on the National Security Implications of Global Climate Change to 2030, the National Intelligence Council (NIC) embarked on a research effort to explore in greater detail the national security implications of climate change in six countries/regions of the world: India, China, Russia, North Africa, Mexico and the Caribbean, and Southeast Asia and the Pacific Island states. For each country/region, we are adopting a three-phase approach.

- In the first phase, contracted research—such as this publication—explores the latest scientific findings on the impact of climate change in the specific region/country.
- In the second phase, a workshop or conference composed of experts from outside the Intelligence Community (IC) will determine if anticipated changes from the effects of climate change will force

inter- and intra-state migrations, cause economic hardship, or result in increased social tensions or state instability within the country/region.

- In the final phase, the NIC Long-Range Analysis Unit (LRAU) will lead an IC effort to identify and summarize for the policy community the anticipated impact on US national security.

This assessment on the impact of Climate Change on Central America and the Caribbean through 2030 is part of the Global Climate Change Research Program contract with the Central Intelligence Agency's Office of the Chief Scientist.

This assessment identifies and summarizes the latest peer-reviewed research related to the impact of climate change on selected countries in Central America and the Caribbean. It draws on the literature summarized in the latest Intergovernmental Panel on Climate Change (IPCC) assessment reports, National Communications to the United Nations Framework (UNFCCC) on Climate Change, statistical data from the United Nations Economic Commission for Latin America and the Caribbean (ECLAC) and on other peer-reviewed research literature and relevant reporting. It includes such impacts as sea level rise, water availability, agricultural shifts, ecological disruptions and species extinctions, infrastructure at risk from extreme weather events (severity and frequency), and disease patterns. This paper addresses the extent to which the countries in the region are vulnerable to impact of climate change. The targeted time frame is to 2030, although various studies referenced in this report have diverse time frames.

This assessment also identifies (Annex B) deficiencies in climate change data that would enhance the IC understanding of potential impacts on Central America and the Caribbean and other countries/regions.

EXECUTIVE SUMMARY

Mexico, the countries of the Caribbean, and Central America examined in this report are at risk from the impacts of climate change in the next 20 years because they will be exposed to a greater range of climate changes and have a relatively weak adaptive capacity when compared to the world at large. Within the region, climate change is evident in increased temperatures, changes in precipitation, and sea level rise—and perhaps in weather variability and natural disaster events. Countries in this report include Belize, Cuba, the

Dominican Republic, Guatemala, Haiti, Honduras, Mexico, Nicaragua, and Panama; Puerto Rico is also discussed.

Steady increases within the region in the number of extreme weather events—hurricanes, storms, and droughts—and their effect on infrastructure, public health, loss of human life and agriculture may be attributable to climate change. The countries reviewed do not yet have a full understanding of the potential impacts of future climatic changes and are not prepared to prevent or reduce those impacts.

Regional leaders are aware of these challenges and have begun to make commitments and agreements that will enhance their understanding of future climate change, their own adaptive capacity, and where critical changes and investments need to be made. Leaders have not addressed the problem from a preventive perspective through policy changes or infrastructure investments because of a lack of systematic analysis that quantifies and qualifies the potential impact to the region, allowing the development of relevant and economically viable options. At present the region is still responding to climate change in a reactive manner.

- Regional leaders realize that leaving the situation "as is" will exacerbate their fragile economies, resources, and adaptive capacity but lack strategic plans to address the issue.
- Most countries in the region are signatories to many multilateral environmental agreements (See Annex C) but are only now beginning to implement such agreements.
- There are significant gaps in the ability to fully understand in a systemic way all the dimensions of climate change impacts at the economic, social, and/or environmental level in the region. There are gaps and deficiencies in data, systematic methodologies/analysis, and tools to monitor, share, and track information and events at the local, national, and regional levels.

Efforts are starting to reduce systemic knowledge gaps. There is insufficient funding by regional governments to undertake detailed modeling that would result in information to rank and evaluate the financial viability of potential climate change adaptation projects. Several entities at the national and regional levels are working to develop improved analytical methods and information sharing as well as better data and data availability.

- In September 2008, the Economic Commission for Latin America and the Caribbean (ECLAC) announced that it would undertake multiple studies to review how climate change is affecting regional economies. Currently, the consensus is that climate change is likely to impose serious economic consequences for the Central American and Caribbean regions, making it increasingly difficult to respond to the challenges of poverty reduction, higher human development, and environmental sustainability linked to the attainment of the United Nations Millennium Development Goals.
- Upcoming studies by the ECLAC are expected to contribute to a better understanding of the economic impact of climate change in the region and will outline the costs and benefits of needed related policy responses, both in terms of mitigation and adaptation.

In this report, information available for a selected set of Mexico, Caribbean, and Central American countries has been reviewed to start understanding the projected climate change variability, given certain scenarios to 2030, as well as to start an initial assessment of these countries' current adaptive capacity to reduce such effects.

Very limited modeling and analysis are available for the countries of interest. Because of that, this initial analysis draws heavily on the respective Governments First National Communication to the Conference of the Parties of the United Nations Framework Convention on Climate Change (UNFCCC). These reports offer the most comprehensive and comparable information available today. In the case of Mexico, the Third Communication was used to review summary impacts. This review, however, was primarily focused on improving inventories of greenhouse gases across all types and production of energy as well as the greenhouse gases generated by major economic activity.

This review identifies the following high-priority risks:

- *Energy*. Energy resources, production, and use vary widely across the countries under review. As all the countries experience population growth, economic growth, and industrialization, they will increase their need and demand for energy. All the countries under review rely on imported fossil fuels, with the exception of Mexico, which is a net exporter of energy resources. In most of the countries, the largest generator of greenhouse gases is the energy sector. Although they are very small contributors to global emissions, most of the countries will benefit from increasing use of renewable energy. Most have begun efforts to evaluate and implement small renewable energy projects, such as solar energy in rural areas of El Salvador, wind energy in Nicaragua and Costa Rica, and an intensive effort in Dominican Republic to evaluate hydro-generated electricity.

- *Agriculture.* The agricultural sector climate related research for most of the countries in this review is limited. Where research is available, productivity losses are projected for optimist, moderate, and pessimist scenarios for some key food crops with estimates that vary from 10 percent to more than 50 percent degradation by the year 2030.
- *Water Resources.* The majority of the population in most of the countries reviewed lives in coastal areas, which are highly vulnerable to severe climate changes. As populations continue to grow in the same areas, increasing water extraction and rising sea levels are expected to have severe impacts on the quantity and quality of water available. Many of these countries' aquifers are open to ocean waters and are already experiencing increased salinity. Rising sea levels will accelerate the deterioration of aquifers and available water resources.
- *Migration.* In Central America, an increase in intra-regional migration during the 1 980s and 1990s as well as extra-regional migration was the result of social unrest and economic contraction. Future patterns of migration are not expected to change significantly. Moreover, the inability of countries in the region to adapt and recover from severe climate events with major impacts on their economies will continue to promote migration outside the region, in particular, to the United States and Canada. The large number of immigrants coming to the United States in the past 20-25 years will facilitate this movement.

Most of the countries under review have submitted their First Communication to the UNFCCC; Mexico has submitted its third. Significant work and analysis needs to be done to fully capture the impact on socio-economic systems and their current ability to recover, adapt, and reduce the effects of climate change.

The great variation of information available for each country reduces the ability to compare the full set of key indicators across all countries in a consistent manner.

INTRODUCTION AND BACKGROUND

Current State of the Region

Mexico, islands in the Caribbean, and the countries of Central America are vulnerable to climate change. Principal components of this vulnerability include their extensive coastlines, current economic dependence on agriculture, the potential for storm damage, scarcity of fresh water, and limited capacity to adapt. This report examines changes in the climate that can be expected, the impacts of those changes on the region and on individual countries, and the resources they can call upon to mitigate or adapt to those impacts. The focus is on ten islands and countries:[1] Belize, Cuba, the Dominican Republic, Guatemala, Haiti, Honduras, Mexico, Nicaragua, Panama, and Puerto Rico.[i] Figure 1 shows the area with the selected countries' names in red.

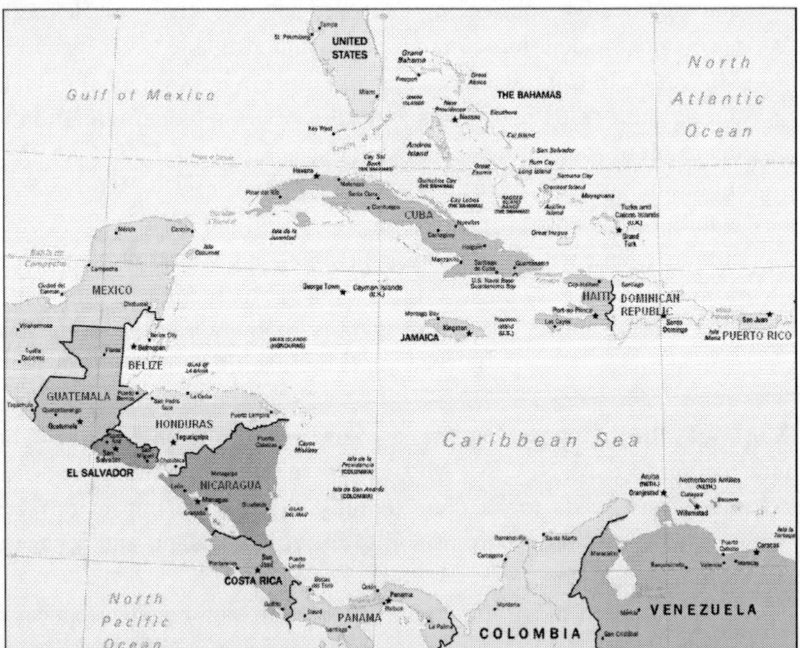

Figure 1. The Central American and Caribbean region with study countries' names in red

[1] Other countries in this region, such as Costa Rica, El Salvador, and Jamaica, as well as Bermuda and other islands, are mentioned in the report but not discussed in detail.

Belize

Belize borders the Caribbean Sea to the east, Mexico to the north, and Guatemala to the west and south. Its total area is 22,966 km^2, including 160 km^2 of water. The country is mostly a flat, swampy coastal plain, with low mountains in the southern portion. It is subject to frequent hurricanes and coastal flooding. Current environmental issues include deforestation, water pollution, and solid and sewage waste disposal. Belize's 2009 population is estimated at 308,000, growing at 2 percent annually (2009 estimate). Life expectancy at birth is 68 years. Fifty percent of the population are Roman Catholic, 27 percent Protestant, 14 percent other religions, and 9 percent claim no religion. Gross Domestic Product (GDP) per capita is $8,600 [US dollar (USD) equivalent; 2008 estimate].

Cuba

The Republic of Cuba is an island between the Caribbean Sea and the North Atlantic Ocean. Its total area is 110,860 km^2 (no areas of water). Cuba's terrain is mostly flat or rolling plains, with hills and mountains in the southeast of the island. It is subject to both hurricanes and droughts. Current environmental issues are air and water pollution, biodiversity loss, and deforestation. The 2009 population is estimated at 11.5 million, with a growth rate of 0.2 percent annually. Life expectancy at birth is 77 years. Religions include Roman Catholicism, Protestantism, Jehovah's Witnesses, Judaism, and Santeria. GDP per capita in 2008 was estimated at $9,500 USD.

Dominican Republic

The Dominican Republic occupies the eastern two-thirds of the island Hispaniola, between the Caribbean Sea and the North Atlantic Ocean. Its area totals 48,730 km^2, including 350 km^2 of water. In the Dominican Republic, highlands and mountains are interspersed with fertile valleys. The country experiences severe storms and hurricanes, occasional flooding, earthquakes, and periodic droughts. Current environmental issues include water shortages, soil erosion and consequent coral reef damage, and deforestation. The 2009 population is estimated at 9.6 million, with a growth rate of 1.5 percent annually (2009 estimate). Life expectancy at birth is 74 years. Citizens are 95 percent Roman Catholic. GDP per capita in 2008 was estimated at $8,100 USD.

Guatemala

The Republic of Guatemala has two coasts: on the Gulf of Honduras to the east and on the North Pacific Ocean to the south. Guatemala borders Mexico and Belize to the north and Honduras and El Salvador to the south. Its area totals 108,890 km^2, including 460 km^2 of water. Its Caribbean coast is susceptible to hurricanes and severe storms. The country is also subject to volcanic activity and earthquakes. Current environmental issues include deforestation in the Peten rainforest, soil erosion, and water pollution. Guatemala's population in 2009 was estimated at about 13 million, growing at a 2 percent per annum rate. Life expectancy at birth is 70 years. Religions include Roman Catholicism, Protestantism, and indigenous Mayan beliefs. GDP per capita for 2008 was estimated at $5,200 USD.

Haiti

The Republic of Haiti is located on the western third of the island Hispaniola, east of the Dominican Republic and bordered by both the North Atlantic Ocean and the Caribbean Sea. Its area totals 27,750 km^2, including 190 km^2 of water. The country is mostly rough and mountainous. Haiti experiences hurricanes, severe storms, occasional flooding and earthquakes, and periodic droughts. Current environmental issues include radical deforestation, soil erosion, and inadequate potable water; although coral reefs exist, little is known about their condition.[ii] Haiti's population in 2009 was estimated at 9 million, with an annual growth rate of 1.8 percent. Life expectancy at birth is 61 years. Citizens are 80 percent Roman Catholic, 16 percent Protestant, 3 percent other religions, and 1 percent no religion. Roughly half the population is reported to practice voodoo. GDP per capita was estimated for 2008 at $1,300 USD.

Honduras

The Republic of Honduras is bordered by the Caribbean Sea to the north, Guatemala and El Salvador to the west, the North Pacific Ocean to the southwest, and Nicaragua to the south. Its area totals 112,090 km^2, including 200 km^2 of water. Honduras is mountainous in the interior, with narrow coastal plains. It experiences frequent but generally mild earthquakes, as well as hurricanes and floods along its Caribbean coast. Current environmental issues include deforestation, land degradation, soil erosion, and water pollution by mining activities. Honduras' population was estimated at almost 8 million in 2009, with a growth rate of 2 percent (2009 estimate). Life expectancy at birth is 69 years. The population is 97 percent Roman Catholic and 3 percent

Protestant. GDP per capita was estimated at $4,400 USD for 2008 with extremely high inequality.

Mexico

The United Mexican States constitute the southernmost country in North America, bordered on the north by the United States, to the east by the Gulf of Mexico and the Caribbean Sea, to the south by Belize and Guatemala, and to the west and south by the North Pacific Ocean. Mexico's area totals 1,972,550 km^2, including 49,510 km^2 of water. Its terrain is diverse: high mountains, low coastal plains, high plateaus, and desert. It experiences tsunamis along the Pacific coast, and hurricanes on all coasts, as well as volcanic activity and earthquakes in the center and south. Current environmental issues include inadequate waste disposal, scarce natural fresh water resources and pollution in existing resources, deforestation, erosion, desertification, land degradation, air pollution, and land subsidence from groundwater depletion. Mexico's estimated population for 2009 is 111 million, growing at an annual rate of 1 percent. Life expectancy at birth is 76 years. The population is 77 percent Roman Catholic, 6 percent Protestant, and 17 percent unspecified. GDP per capita was estimated at $14,200 USD for 2008.

Nicaragua

The Republic of Nicaragua is situated between Honduras and Costa Rica to the north and south, respectively, and between the North Pacific Ocean and the Caribbean Sea to the west and east, respectively. Its area totals 129,494 km^2, including 9,240 km^2 of water area. Extensive Atlantic coastal plains rise to central interior mountains; the narrow Pacific coastal plain has volcanoes. Nicaragua experiences earthquakes, volcanic activity, landslides, and hurricanes. Current environmental issues include deforestation, soil erosion, and water pollution. The population estimate for 2009 was about 6 million, growing at an annual rate of 1.8 percent. Life expectancy at birth is 69 years. Citizens are 59 percent Roman Catholic, 22 percent Evangelical, 1.6 percent Moravian, 1 percent Jehovah's Witnesses, and 16 percent no religion. GDP per capita was estimated for 2008 at $2,900 USD.

Panama

The Republic of Panama is located on the isthmus between North America and South America, bordered by the Caribbean Sea and the Pacific Ocean. The southernmost country of Central America, Panama sits between Costa Rica and Columbia. Its total area is 78,200 km^2, of which 2,210 km^2 are water. In its center is a line of mountains, with plains and rolling hills in the coastal areas. Toward Columbia is dense jungle, which, combined with forest protections, causes a break in the Pan American Highway; this area is subject to occasional severe storms and forest fires. In its center is the Panama Canal. Current environmental issues include agricultural runoff that pollutes water and threatens fisheries; deforestation; land degradation and soil erosion (with resulting siltation of the Panama Canal); urban air pollution; and environmental degradation caused by extensive mining. The 2009 estimated population is 3.3 million, growing at 1.5 percent annually. Life expectancy at birth is approximately 77 years. The population is 85 percent Roman Catholic, 15 percent Protestant. The service sector is 80 percent of Panama's economy; per capita GDP is $11,600 USD (2008 estimate). The country's growth rate has been above 8 percent in recent years, but both the unemployment rate and inequality in per capita GDP are high.

Puerto Rico

The Commonwealth of Puerto Rico, a self-governing territory of the United States, consists of several islands situated east of the Dominican Republic and west of the Virgin Islands. Its area is 13,790 km^2, including 4,900 km^2 of water. The main island, Puerto Rico, is mostly mountainous but has large coastal areas both in the north and in the south. As all the countries covered in this report, Puerto Rico is subject to hurricanes. Current environmental issues include erosion and occasional droughts with accompanying water shortages. Its population numbers about 4 million (2009 estimate), with a growth rate of 0.3 percent. Life expectancy is 79 years. Roman Catholicism dominates (85 percent), but Protestant, Jewish, indigenous, and African religions are also espoused. Puerto Rico's per capita GDP is $17,800 USD (2008 estimate), and economic activities are largely services and industry.

Emissions

Latin America (the Caribbean, Central, and South America) is responsible for only a small fraction of global carbon emissions (Figure 2). Within the Latin American and Caribbean region, Meso-America—typically thought of as covering some of Mexico south to Honduras and Nicaragua—represents half of the carbon dioxide emissions of Latin America and the Caribbean accounts for less than 15 percent (Figure 3). This figure illustrates the wide variation of carbon dioxide (CO_2) emissions in the region. The highest and most quickly increasing—40 percent between 1990 and 2000—amount comes from South

America, while the lowest and relatively more slowly rising amount comes from the Caribbean.

Although the region is a very small contributor to total worldwide carbon dioxide emissions, the impacts of climate change in this region are already being felt. Temperature increases in the atmosphere and sea, instability in rainfall, and rising sea levels are affecting food production, infrastructure, livelihood, and the health of populations. Extreme weather events (droughts, hurricanes, floods, etc.) have added more stress on an already weakened environment and further eroded the ability of the environment to mitigate their harmful effects.

Economic growth and emissions have moved roughly in the same direction. As developing economies continue urbanization and industrialization, the risk of growing emissions increases because of energy use mix and the inability of economies to become more energy efficient. In the region under evaluation, there has been a wide variety of energy intensity of GDP over the past 37 years. Most of the countries have become more energy efficient, with the exception of Haiti and Nicaragua, two of the lowest-performing countries by many measures. These two countries have gone through many years of political unrest, resulting in economic contraction, capital flight, migration of the best human capital, and inefficiencies at every level of economic activity (see Table 1).

Table 1. Energy Intensity of Gross Domestic Product (2000 Prices = 100) (in thousands of barrels of oil equivalent for US $1 M of GDP).

Country	1970	1980	1990	2000	2007
Costa Rica	1.61	1.36	1.33	1.06	1.17
Cuba	…	…	2.64	1.99	1.86
Guatemala	2.83	2.38	2.71	2.68	2.32
Haití	3.75	3.13	2.42	3.43	4.58
Honduras	3.85	3.18	3.24	2.78	2.64
Mexico	1.14	1.20	1.31	1.13	1.08
Nicaragua	2.38	2.99	3.76	3.88	4.31
Dominican Republic	2.41	1.65	1.47	1.66	1.17
Panama	1.40	1.29	1.21	1.17	1.42
Latin America and the Caribbean	**1.59**	**1.47**	**1.60**	**1.53**	**1.46**

Source: CEPAL/ECLAC [Comisión Económica para América Latina y el Caribe/Economic Commission for Latin America and the Caribbean], *Anuario Estadístico de América Latina y el Caribe: 2008* (United Nations 2009)

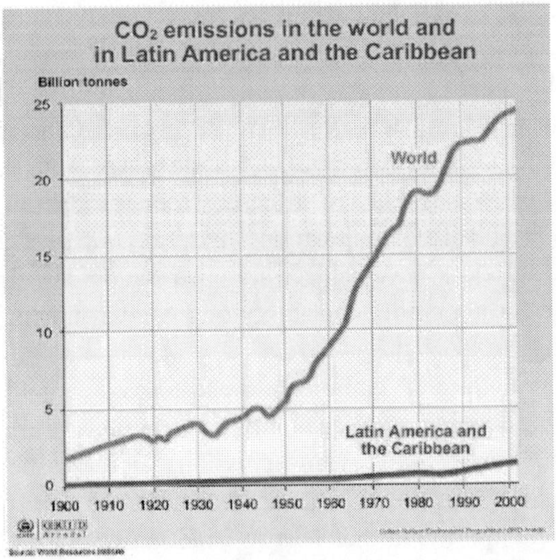

Figure 2. Relative CO_2 emissions in Latin America and the Caribbean. Source: United Nations Environment Program, "Vital Climate Graphics for Latin America and the Caribbean," (UNEP 2003) http://www.grida.no/publications/vg/lac/

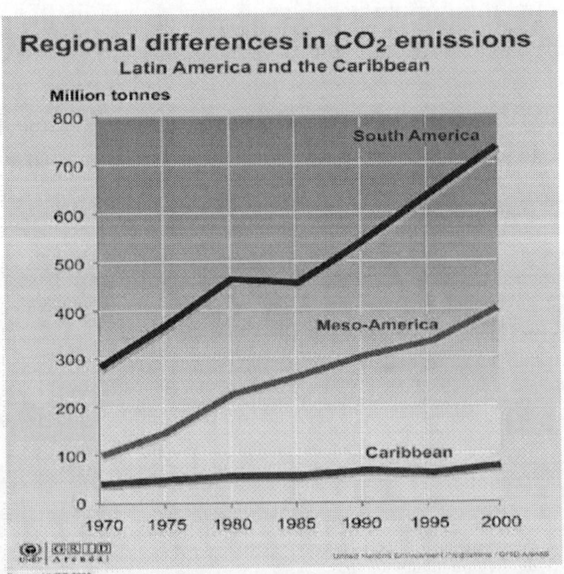

Figure 3. Regional differences in CO_2 emissions. Source: United Nations Environment Program (UNEP), "Vital Climate Graphics for Latin America and the Caribbean," (2003) http://www.grida.no/publications/vg/lac/

Economic Growth and Development

Central America, Mexico, and the Caribbean countries all continue to experience population growth, albeit at different rates, leading to an increase in food demand. Most of the countries in these regions depend greatly on agricultural production. Variations in crop yields, food crops, and cash crops present major food security challenges.

Since 1990, the countries in the region have experienced large disparities GDP. Some have suffered from economic contraction due to political unrest, capital flight, migration of the better- educated segment of the population, and the loss of foreign investments. Examples include Guatemala, El Salvador, Nicaragua, and Haiti from the late 1970s through the 1990s. The sociopolitical challenges of the 1980s and increases in extreme weather events in the 1990s hurt the fragile economies of the region. The absence of a strong legal foundation has also greatly reduced the opportunity for recovery. El Salvador, Guatemala, and Nicaragua were directly affected by insurgencies and increased weather-related natural disasters. At the same time, neighboring countries had to cope with an increase in refugees because of the difficulties associated with war and natural disasters. All these countries have been severely affected by hurricanes, floods, and tropical storms in the past two decades. Regional GDP has shown the effects of all these events through wide fluctuations from one year to the next (see Figure 4).

Energy Systems

The countries in the region selected for evaluation, have mostly fossil fuel-based economies and are mostly net importers of energy. Since 1984 they have continued to increase their overall energy consumption. Except Mexico, primary and secondary energy production has remained below total annual consumption (Figure 5). Primary energy production is the production of energy found in its natural state—wood, natural gas, bagasse,[2] and hydroelectricity. It also includes the amount of fuel extracted and the energy consumed in the production process and the supply to energy producers and conversion. Secondary energy production is derived from the conversion of

[2] Sugarcane fiber left over after juice extraction.

primary energy products. Petroleum, for example, is refined into kerosene and diesel.

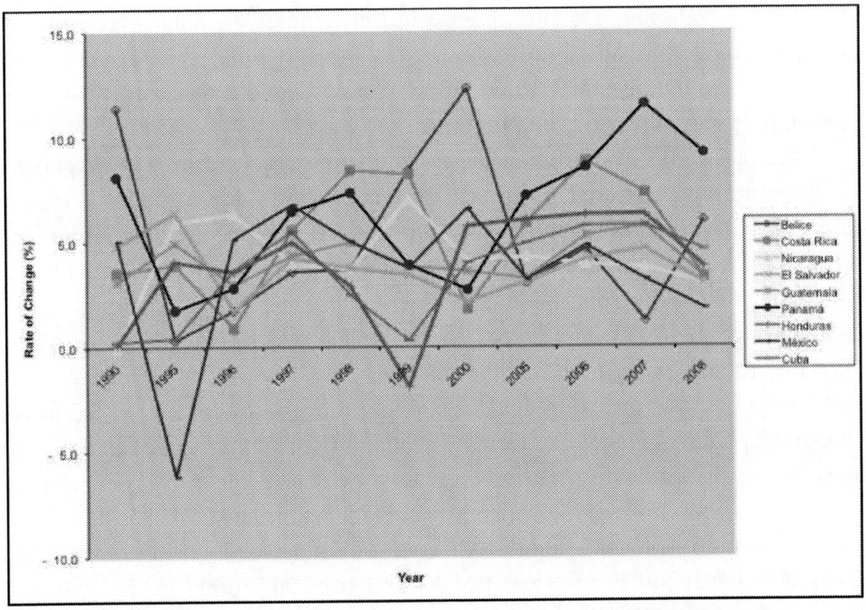

Figure 4. Rate of Change in Gross Domestic Product (GDP) (1990-2008). Data for Haiti and the Dominican Republic are not included. Source: CEPAL/ECLAC [Comisión Económica para América Latina y el Caribe/Economic Commission for Latin America and the Caribbean], *Anuario Estadístico de América Latina y el Caribe: 2008* (United Nations 2009)

As economies industrialize, most countries in this review will remain highly vulnerable to the fluctuations in the cost of oil. Mexico is the only country in the group that is a net exporter of energy resources; all others in this study are net importers of petroleum-based products. During the 1990-2007 time period, regional energy consumption increased 158 percent; in Costa Rica, Nicaragua, and Dominican Republic it increased about 200 percent; and in Panama by 288 percent. Energy consumption is expected to increase as population and economies continue to grow. Figures 6 and 7 illustrate total energy consumption and total energy supply by country and type, respectively. Note that energy supply information is not available for the same time period as that for energy consumption.

Energy supply composition across the countries reviewed remains predominantly based on petroleum, except for Haiti, Nicaragua and Honduras.

These three countries had the lowest annual GDP growth rates within the group from 1990 to 2007. On the other hand, Costa Rica, Cuba, Panama, and Dominican Republic have the largest shares of oil-based energy and experienced the largest annual GDP growth rates.

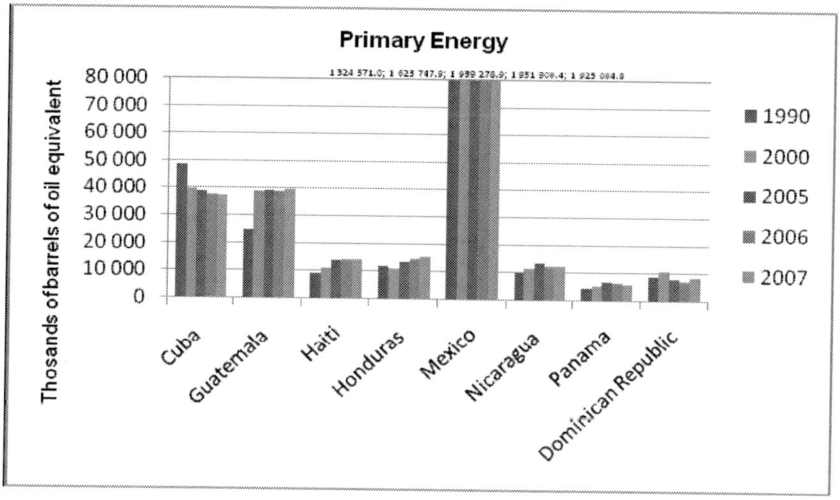

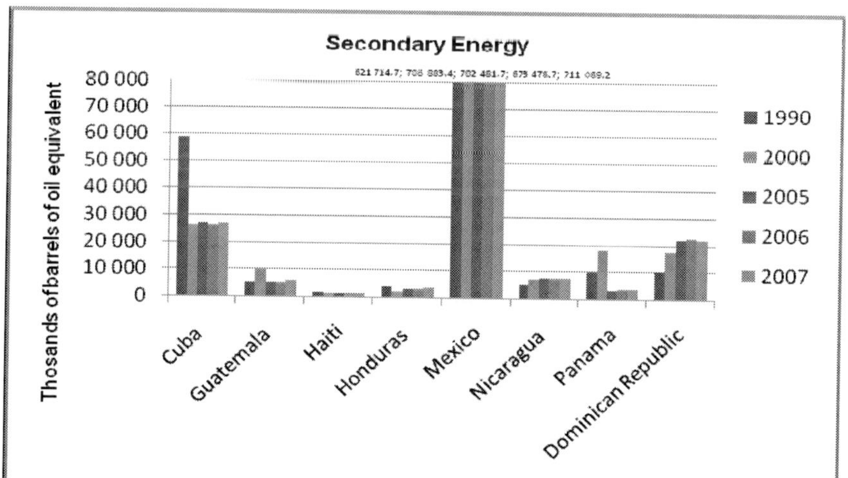

Figure 5. Primary and Secondary Energy Production by Country and Regions. Note: no data for Belize and Puerto Rico. Source: CEPAL/ECLAC [Comisión Económica para América Latina y el Caribe/Economic Commission for Latin America and the Caribbean], *Anuario Estadístico de América Latina y el Caribe: 2008* (United Nations 2009)

As noted earlier, all countries except Mexico are net importers of petroleum-based products. In the Dominican Republic oil-based energy supply remains significant and accounted for 74 percent of total energy in 2005 and 79 percent in 2002. The island nations of Cuba, Puerto Rico, Haiti, and the Dominican Republic remain particularly vulnerable to supply of petroleum-based energy products since they must be brought by ship to the islands for refining and processing. Hydroenergy plays a significant role only in Costa Rica, where it accounted for 18-24 percent of supply; for the other countries it ranged from 0.1 to 9.8 percent for Cuba and Panama respectively.

Food Production and Drinking Water Supply

Central America, the Caribbean, and Mexico have economies with significant agricultural sectors though agricultural land use as part of total land area varies widely. In Belize only six percent of the total land area was devoted to agriculture in 2005 reflecting the fact that over 50 percent of GDP comes from the services industry, particularly tourism. The comparable figure for the Dominican Republic was 70 percent, Costa Rica and Haiti 57 percent, Cuba 60 percent, and Mexico 55 percent (Figure 8). All the countries reviewed have maintained relatively stable ratios of agricultural land use to total land area for the past 27 years.

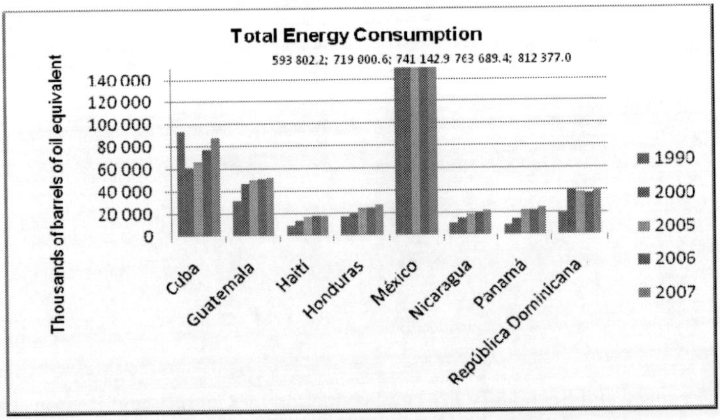

Figure 6. Consumption by country. Note: no data for Belize and Puerto Rico. Source: CEPAL/ECLAC [Comisión Económica para América Latina y el Caribe/Economic Commission for Latin America and the Caribbean], *Anuario Estadístico de América Latina y el Caribe: 2008* (United Nations 2009)

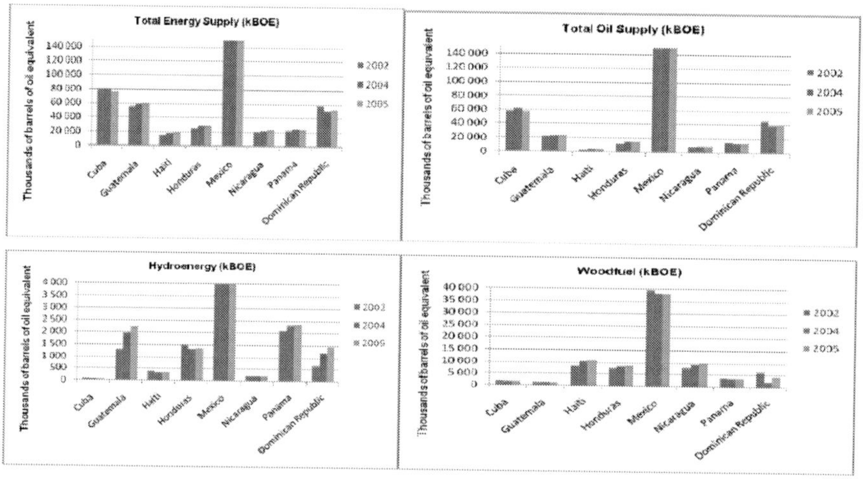

Figure 7. Energy supply by type (2002-2005). Note: no data for Belize and Puerto Rico. Source: CEPAL/ECLAC [Comisión Económica para América Latina y el Caribe/Economic Commission for Latin America and the Caribbean], *Anuario Estadístico de América Latina y el Caribe: 2008* (United Nations 2009)

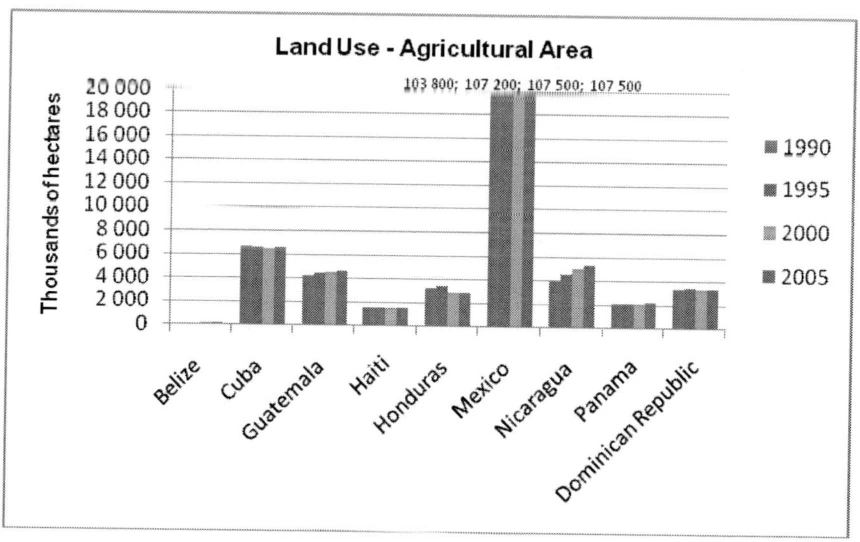

Figure 8. Agricultural area and total land area by country in hectares. Source: CEPAL/ECLAC [Comisión Económica para América Latina y el Caribe/Economic Commission for Latin America and the Caribbean], *Anuario Estadístico de América Latina y el Caribe: 2008* (United Nations 2009)

Table 2. Percentage of total population living in poverty by country

POOR AND INDIGENT POPULATION, URBAN AND RURAL AREAS (Percentage of total population)									
Country		Poverty				Extreme poverty			
	Years	Total	Urban	Rest	Rural	Total	Urban	Rest	Rural
Guatemala	1998	49.1	...	...	69.0	16.0	...	...	41.8
	2002	45.3	...	...	68.0	18.1	...	...	37.6
	2006	42.0	...	...	66.5	14.8	...	...	42.2
Honduras	1994	74.5	68.7	80.4	80.5	46.0	38.3	53.7	59.8
	1999	71.7	64.4	78.8	86.3	42.9	33.7	51.9	68.0
	2007	59.9	47.8	64.0	78.8	26.2	18.0	32.5	61.7
Mexico	1994	36.8	...	...	56.5	9.0	...	...	27.5
	2000	32.3	...	...	54.7	6.6	...	...	28.5
	2006	26.8	...	...	40.1	4.4	...	...	16.1
Nicaragua	1993	66.3	58.3	73.0	82.7	36.8	29.5	43.0	62.8
	2001	63.8	50.8	72.1	77.0	33.4	24.5	39.1	55.1
	2005	54.4	48.7	58.1	71.5	20.8	16.4	23.7	46.1
Panama	1994	25.3	...	...	...	7.8	...	...	...
	1999	20.8	...	...	...	5.9	...	...	...
	2007	18.7	...	...	46.6	5.0	...	...	24.1
Dominican Republic	2002	42.4	...	...	55.9	16.5	...	...	28.6
	2006	41.8	...	...	49.5	18.5	...	...	28.5
	2007	43.0	...	...	47.3	19.0	...	...	24.6
Latin America	1994	38.7	...	...	65.1	13.6	...	...	40.8
	2000	35.9	...	...	62.5	11.7	...	...	37.8
	2007	28.9	...	...	52.1	8.1	...	...	28.1

Data not available for Cuba, Belize, Haiti, and Puerto Rico

Note: no data for Belize, Cuba, Haiti, and Puerto Rico.

Source: CEPAL/ECLAC [Comisión Económica para América Latina y el Caribe/Economic Commission for Latin America and the Caribbean], *Anuario Estadístico de América Latina y el Caribe: 2008* (United Nations 2009)

Although the areas dedicated to agricultural activity and food production are significant in almost all the countries studied, a large portion of the population lives in poverty and struggles to survive. Table 2 shows the percentage of the population living in poverty and extreme poverty. Those with income amounting to less than twice the cost of a basic food basket[3] are considered to be living in poverty. Those with income amounting to less than

[3] The food basket is a concept used in poverty measurement; it differs in components by country or region according to local diets and availability but must provide adequate calories and protein. Traditionally, a food basket has represented the minimum food items required for a family over a one-month period.

the cost of a basic food basket are considered to be living in extreme poverty. Costa Rica and Panama are the only two countries of those for which we have information that have less than 20 percent in poverty and no more than 5 percent in extreme poverty. Figure 9 shows the Consumer Price Index (CPI)—the change in the cost of the food basket—with the base year of 2000.

The countries under review have experienced a steady increase in CPI that has translated into reduced access to the basic food basket. By 2008, Haiti had the highest Index (350) followed by the Dominican Republic (290), Nicaragua (202) and Honduras (188). These countries also have been affected by severe climate variations since the 1990s and highly variable inflation rates. Although food production indexes have remained positive (Figure 10), in Nicaragua, Honduras, and Guatemala a significant portion of the population has experienced a steady decline in access to food because of reduced purchasing power.

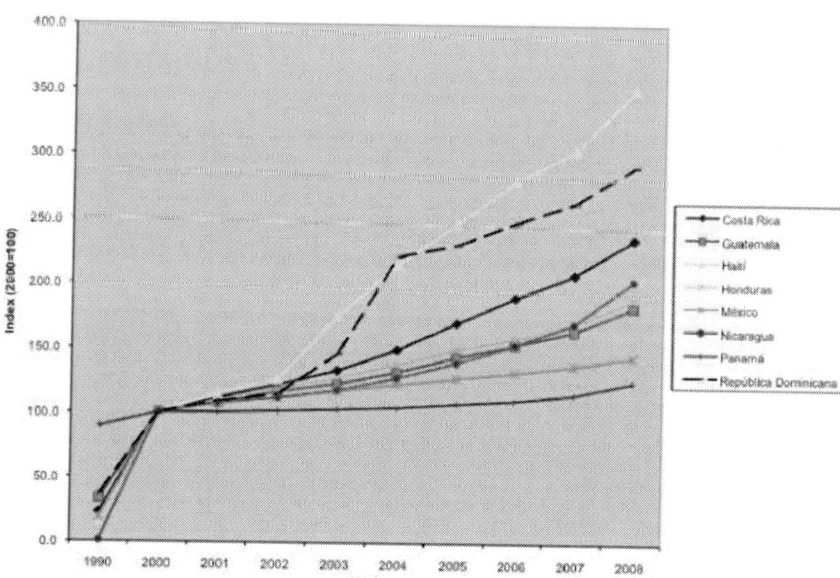

Figure 9. Consumer Price Index by Country. (2000=100). Source: CEPAL/ECLAC [Comisión Económica para América Latina y el Caribe/Economic Commission for Latin America and the Caribbean], *Anuario Estadístico de América Latina y el Caribe: 2008* (United Nations 2009)

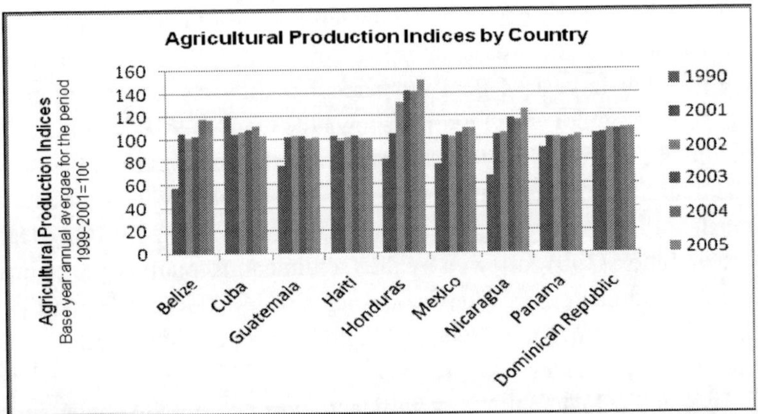

Source: CEPAL/ECLAC [Comisión Económica para América Latina y el Caribe/Economic Commission for Latin America and the Caribbean], Anuario Estadístico de América Latina y el Caribe: 2008 (United Nations 2009).

Figure 10. Food Production Indexes by Country

Table 3. Basic services supplied in six countries

Country	Area	Piped water		Excreta disposal system		Electric lighting	
		1995 c/	2007 c/	1995 c/	2007 c/	1995 c/	2007 c/
Costa Rica	Total	...	95.2	...	25.6	...	99.1
	Urban	...	99.6	...	39.3	...	99.8
	Rural	...	88.5	...	5.0	...	98.0
Guatemala	Total	63.6	76.3	32.6	40.3	64.1	81.8
	Urban	89.6	90.0	73.3	68.4	91.2	93.7
	Rural	43.6	60.6	1.4	7.6	43.4	68.0
Honduras	Total	70.7	82.5	26.8	33.0	55.3	73.9
	Urban	80.6	93.6	51.5	62.9	86.1	97.9
	Rural	62.1	71.8	5.6	4.1	28.9	50.7
Mexico	Total	84.3	...	60.8	73.5	95.9	98.5
	Urban	94.1	...	81.7	90.0	99.3	99.7
	Rural	67.9	...	25.8	42.1	90.5	96.1
Nicaragua	Total	61.0	...	61.1	26.4	69.3	73.9
	Urban	86.0	...	56.7	21.1	90.8	95.5
	Rural	27.0	...	67.0	33.9	40.6	43.7
Dominican Republic	Total	70.1	71.9	19.8	23.2	88.5	...
	Urban	82.8	80.6	30.5	32.3	...	100.0
	Rural	48.1	55.4	1.2	5.8	...	68.6

Note: no data for Belize, Haiti, Panama, and Puerto Rico.

Source: CEPAL/ECLAC [Comisión Económica para América Latina y el Caribe/Economic Commission for Latin America and the Caribbean], *Anuario Estadístico de América Latina y el Caribe: 2008* (United Nations 2009).

Table 3 shows the extent to which six countries provided their citizens with basic services, such as drinking water, sanitation, and electricity. Drinking water across the selected countries varies significantly between urban areas and rural areas. In 2007, Costa Rica had drinking water supply services available for 99.2 percent of total urban dwellings and 88.5 percent of total rural dwellings. The respective figures in Guatemala were 90 percent and 60 percent, and in the Dominican Republic 80.6 and 55.4 percent. Among the countries that provided information, Nicaragua has the lowest percentage of population with available basic services.

Human Health

Since 1990 the region has experienced a series of re-emerging diseases following such severe climatic events as floods, hurricanes, and droughts. Evidence points to increases in several communicable diseases, such as dengue, malaria, Hantavirus pulmonary syndrome, and the reemergence of a large host of infectious diseases following years in which there were El Niño/Southern Oscillation (ENSO) events.

Projected Regional Climate Change

Central American and Caribbean countries span the tropics and the subtropics and include continental land masses, island chains, and mountain ranges of varying orientations and elevations.[iii] The general climate of the region is described as dry winter/wet summer. The temperature range within the region is small due to its maritime tropical characteristics except for the mountainous areas where temperatures are modulated by changes in altitude.

Rainfall intensity and timing determines the climate classification and the meteorological features of the climate in the region. The topography of countries with significant mountains can influence variations in annual rainfall, the timing of peak rainfall, and the length of the rainy season. Windward slopes of the larger mountainous islands tend to have the highest amounts of rainfall.

The continental landmass of Central America lies between two oceans and contains some of the most diverse coastal and marine ecosystems in the world. Tropical forests, particularly in Costa Rica, are a significant sink for greenhouse gases and are of great value to countries interested in gaining credits under trading mechanisms such as those specified in the Kyoto protocol to the UNFCCC.[4]

The Caribbean experiences a wet season from May through October and a dry season from November through April. During late July or early August, a short-lived dry period may occur. In the winter and early summer, the occasional intrusion of a mid-latitude polar front can influence weather patterns by bringing cool, moist air to the region.

Tropical storms and hurricanes are a perennial feature of the Caribbean. The official hurricane season lasts from June 1 to November 30. The phase of the ENSO influences the likelihood of hurricane formation in the Atlantic. During El Niño, (the ENSO warm phase), the formation of tropical hurricanes in the Atlantic is inhibited. Alternately, during La Niña (the ENSO cold phase) the formation of hurricanes is enhanced.

The Atlantic Multidecadal Oscillation (AMO) is another important natural influence on air temperatures, precipitation levels, and storm activity in the Caribbean. The AMO has cool and warm phases, each lasting several decades. The phase of the AMO also plays a role in suppressing tropical storm formation. The AMO and ENSO are several features of the Caribbean climate that can complicate the observations of temperature and precipitation trends.

Climate Observations

Evidence of intensified climate variability can be seen in multiple key economic, social, and environmental indicators. A review of regional natural disasters that are weather-related demonstrates that the frequency and impact of severe events has steadily increased in both number and affected population.

Since 1990 the Central American and Caribbean region has experienced a steady rise in the number of people affected by severe events - floods, hurricanes, and storms (Table 4). The increase has occurred because most urban centers are located in the coastal areas.

---

[4] For more information on trading mechanisms specified in the Kyoto protocol, see http://unfccc. int/Kyoto_protocol/items/2830.php.

Table 4. Climate-related natural disasters in Latin America and Caribbean Region (1990–2006)

		1990	1995	1996	1997	1998	1999	2000	2001	2002	2003	2004	2005	2006
Floods														
Caribbean c/														
	Number of Events	3	2	6	1	0	1	3	1	4	6	2	5	6
	People Afected	5,000	35,780	39,066	35,000	0	0	675	5,070	100,085	216,060	39,040	290,302	91,190
	Deaths (Human)	4	0	6	0	0	13	18	26	43	72	3,353	54	17
Latin America d/														
	Number of Events	15	13	6	11	8	20	28	22	31	27	16	21	14
	People Afected	207,236	326,185	310,000	970,207	436,300	1,904,352	307,866	711,782	731,969	409,955	587,016	703,735	643,555
	Deaths (Human)	173	298	40	700	464	30,852	469	225	307	452	249	434	244
Latin America & Caribbean e/														
	Number of Events	18	15	12	12	8	21	31	23	35	33	18	26	20
	People Afected	212,236	361,965	349,066	1,005,207	436,300	1,904,352	308,541	716,852	832,054	626,015	626,056	994,037	734,745
	Deaths (Human)	177	298	46	700	464	30,865	487	251	350	524	3,602	488	261
Landslides														
Caribbean c/														
	Number of Events	0	0	1	0	0	0	1	0	0	0	1	0	0
	People Afected	0	0	175	0	0	0	0	0	0	0	1,200	0	0
	Deaths (Human)	0	0	0	0	0	0	10	0	0	0	2	0	0
Latin America d/														
	Number of Events	1	4	5	2	6	4	5	3	5	4	2	2	2
	People Afected	0	0	7,000	30,000	2,600	200	350	0	0	1,810	5,751	2,500	0
	Deaths (Human)	33	165	216	312	272	112	122	114	285	136	40	70	21
Latin America & Caribbean e/														
	Number of Events	1	4	6	2	6	4	6	3	5	4	3	2	2
	People Afected	0	0	7,175	30,000	2,600	200	350	0	0	1,810	6,951	2,500	0
	Deaths (Human)	33	165	216	312	272	112	132	114	285	136	42	70	21
Droughts														
Caribbean c/														
	Number of Events	0	0	0	1	1	0	2	0	0	1	1	0	0
	People Afected	0	0	0	607,200	820,000	0	0	0	0	35,000	0	0	0
	Deaths (Human)	0	0	0	0	0	0	0	0	0	0	0	0	0
Latin America d/														
	Number of Events	2	1	1	3	4	3	3	5	4	1	4	2	1
	People Afected	2,483,160	0	0	324,000	10,100,000	105,000	21,125	1,896,596	103,500	0	192,500	52,990	0
	Deaths (Human)	0	0	0	0	0	12	0	41	0	0	0	0	0
Latin America & Caribbean e/														
	Number of Events	2	1	1	4	5	3	5	5	4	2	5	2	1
	People Afected	2,483,160	0	0	931,200	10,920,000	105,000	21,125	1,896,596	103,500	35,000	192,500	52,990	0
	Deaths (Human)	0	0	0	0	0	12	0	41	0	0	0	0	0
Hurricanes/Tornados/Tropical Storms														
Caribbean c/														
	Number of Events	6	7	4	0	6	9	1	8	8	2	17	17	2
	People Afected	2,000	70,260	294,995	0	1,084,000	251,857	62,000	5,920,175	327,720	10,000	943,601	2,642,816	15,260
	Deaths (Human)	0	16	63	0	550	6	14	42	15	34	2,836	111	5
Latin America d/														
	Number of Events	1	8	7	9	10	2	8	14	11	3	4	19	3
	People Afected	0	30,062	727,724	757,405	3,189,660	2,000	32,910	100,452	616,667	9,900	151,845	3,332,649	270,700
	Deaths (Human)	38	194	122	287	19,045	10	35	90	104	16	29	1,755	15
Latin America & Caribbean e/														
	Number of Events	7	15	11	9	16	11	9	22	19	5	21	36	5
	People Afected	2,000	100,322	1,022,719	757,405	4,273,660	253,857	94,910	6,020,627	944,387	19,900	1,095,446	5,975,465	285,960
	Deaths (Human)	38	210	185	287	19,595	16	49	132	119	50	2,865	1,866	20

Note: Latin American includes South America as well as Central America.

Source: CEPAL/ECLAC [Comisión Económica para América Latina y el Caribe/Economic Commission for Latin America and the Caribbean], *Anuario Estadístico de América Latina y el Caribe: 2008* (United Nations 2009).

In 1998 Hurricane Mitch was one of six hurricanes that caused significant damage in the region. The countries affected still have not fully recovered from the disaster. In Honduras, at least 90 percent of the population was without water; in Nicaragua, 32 percent of the water infrastructure was damaged; and in Guatemala, the water and sewage systems in 396 communities were damaged and 20,000 latrines destroyed.

The Central American and Caribbean regions have followed the global trend of warming surface temperatures that the rest of the world has experienced. Some experts believe a warming climate may contribute to an increase in frequency and intensity of the ENSO phenomenon.

Warmer-than-average temperatures in the Pacific around the equator reduce the normal difference in the sea surface temperature between the Pacific's eastern and western sides, affecting wind patterns. At the same time, the warmer waters move toward the east along the equator, while the weakened trade winds reduce the equatorial Pacific's capacity to absorb cold water, thus consolidating the temperature anomaly. This affects the patterns that warm the atmosphere. It also affects wind direction, sea currents, and storm patterns.[iv]

In Central America ENSO leads to excessive rainfall along the coast of the Atlantic Ocean, while the Pacific coast remains dry. The effects of ENSO have caused large increases in rainfall in some areas and extended droughts in others. There was a high incidence of hurricanes and tropical storms in 1998, which was a key year for ENSO effects in the warming of ocean surface water. Figure 11 shows climate impacts and the areas affected by above-normal surface ocean temperature in Mexico, Central America, and South America during 1998.

Shortcomings in the frequency and quality of past climate data in the region present a problem in accurately assessing trends. In 2001, a workshop was held to develop climate indices for the Caribbean region.[v] At that time the region had significant problems in digitizing and developing quality assurance methods for daily weather data. Jamaica was the only country in the region to have developed a digital archive, and that was lost in a fire in 1992. Data from 30 stations were used during the 2001 exercise, primarily stations in the Caribbean islands, with one coastal Florida station, and 4 stations from Belize. The results showed that over the last few decades the number of very warm days and nights has dramatically increased and the number of very cool days and nights has decreased. The maximum number of consecutive dry days has also decreased, but the number of heavy rainfall events has increased.

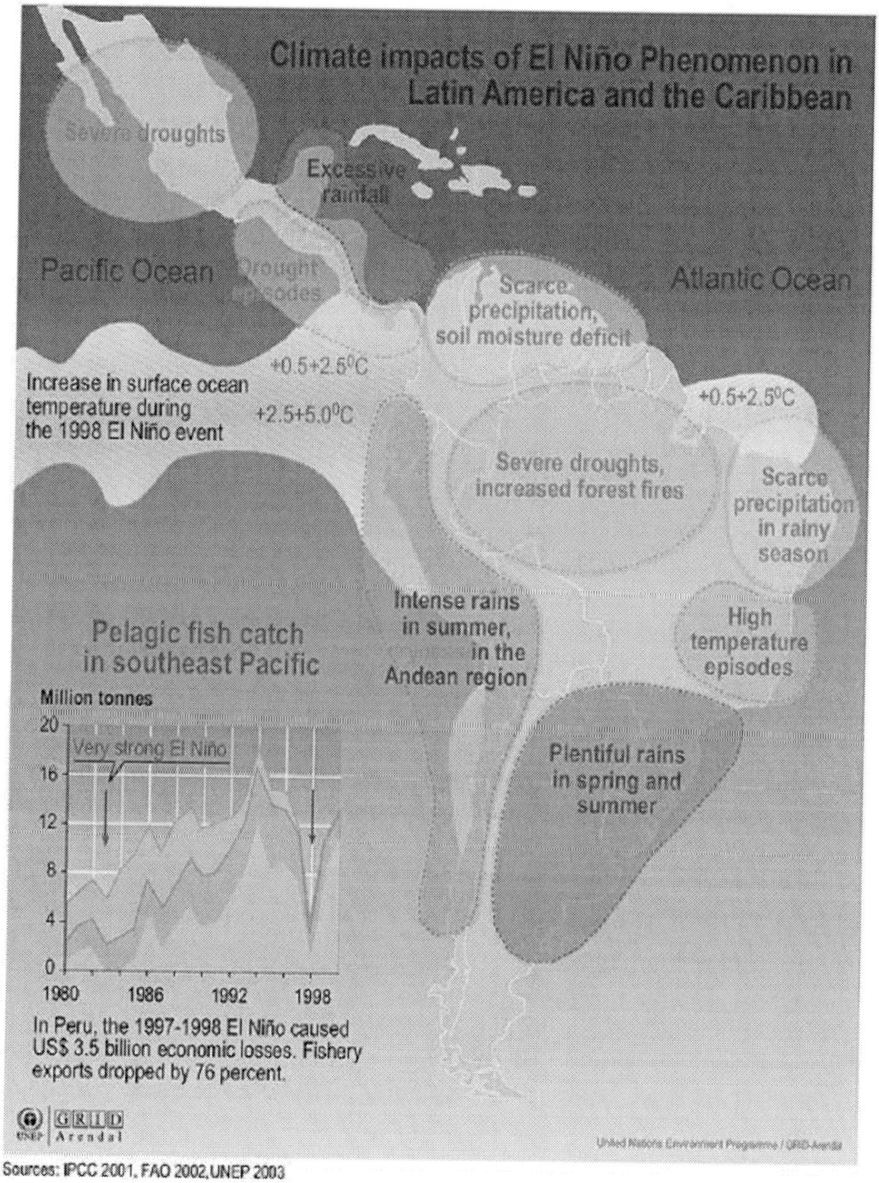

Source: United Nations Environment Program, "Vital Climate Graphics for Latin America and the Caribbean," (UNEP 2003) http://www.grida.no/publications/vg/lac/page/2753.aspx.

Figure 11. Impact of El Niño in Latin America and the Caribbean

The 1998 IPCC[vi] assessment reported that on average the Caribbean islands experienced an increase in temperature exceeding 0.5°C from the year 1900 until the time of the report. Over the same period there had been a significant increase in rainfall variability, with mean annual total rainfall declining by approximately 250 mm. However the decreasing rainfall trend was not significant.

The most recent IPCC assessment[vii] reports that air temperatures in the Caribbean have been increasing by as much as 0.1°C per decade and sea levels have been increasing by approximately 2 mm per year over the last few decades.

Data show that there is currently a significant drying trend in the Caribbean and CentralAmerican.[viii] These include a satellite estimate since 1979 and several land-based observational data sets. A multi-model ensemble mean prediction of precipitation change in the region suggests this drying trend is likely to continue. Intermodel agreement on the amplitude of the drying trend yields median amplitude of between 0.5 and 1 mm per day, per 100 years over most of the region.

In the Commonwealth of the Bahamas the data show that mean daily maximum temperatures for July have increased at the rate of 3.6ºF (2°C) per 100 years and more recently at the rate of 4.8ºF (2.6°C) per 100 years.[ix] Sea level rise is expected to occur at a rate of 0.06 inches (1.5 mm) per year, with a sea level rise of about 8 inches (20 cm) by 2060. Observations taken in neighboring islands suggest that rises of 6 to 10 inches (15.2 to 25.4 cm) per 100 years can be expected.[x]

Ecological changes in Central American have substantiated the influence of climate change. For example, vegetation changes have been observed in the tropical montane cloud forests of Costa Rica. The changes suggest that atmospheric warming has raised the average altitude of the base of the orographic cloud bank during the dry season.[xi] Changes in populations of birds, lizards, and anurans[5] all reflect a broad response to regional climate change that includes widespread amphibian extinctions in remote highland forests.

[5] An order of animals in the class Amphibia that includes frogs and toads.

Climate Predictions (Modeling)

Although Global Circulation (or Climate) Models (GCMs) can be used to infer climate changes in specific regions, it is far preferable to develop models that have a resolution sufficient to resolve local and regional scale changes. There are many challenges in reliably simulating and attributing observed temperature changes at regional and local scales. At these scales, it is hard to identify long-term changes expected from external forcings because of the large natural climate variability.

The procedure of estimating the response at local scales based on results predicted at larger scales is known as "downscaling." The two main methods for deriving information about the local climate are (1) dynamical downscaling (also referred to as "nested modeling" using "regional climate models" or "limited area models") and (2) statistical downscaling (also referred to as "empirical" or "statistical-empirical" downscaling). Chemical composition models include the emission of gases and particles as inputs and simulate their chemical interactions; global transport by winds, and removal by rain, snow, and deposition to the earth's surface.

Downscaled regional climate models rely on global models to provide boundary conditions and the radiative effect of well-mixed greenhouse gases for the region to be modeled. There are three primary approaches to numerical downscaling: (1) limited-area models, (2) stretched-grid models, and (3) uniformly high resolution atmospheric GCMs (AGCMs) or coupled atmosphere- ocean (-sea ice) GCMs (AOGCMs).

GCMs simulate changes in climate under scenarios of future greenhouse gas and aerosol emissions. The 2000 IPCC Special Report on Emission Scenarios (SRES)[xii] laid out the four basic scenario families used by IPCC scientists to predict future climate change; they are summarized in Table 5. This set of scenarios is designed to represent the range of possible future global conditions that will influence greenhouse gas emissions. The scenarios are based on consistent and reproducible assumptions about global forces that affect greenhouse gas emissions, including economic development, population, and technological change.

The magnitudes and patterns of the projected rainfall changes differ significantly among models, probably due to their coarse resolution. The Atlantic and Pacific Oceans are strongly influenced by natural variability occurring at 10-year intervals, but the Indian Ocean appears to be exhibiting a steady warming. Natural variability (from ENSO, for example) in ocean-atmosphere dynamics can lead to important differences in regional rates of surface-ocean warming that affect the atmospheric circulation and hence warming over land surfaces.

Table 5. Summary of IPCC emissions scenarios

Emission Scenario	Economic Development	Global Population	Technology Changes	Theme
A1	Very rapid	Peaks around mid-21st century and declines thereafter	Rapid introduction of new and more efficient technologies	Convergence among regions; increased cultural and social interactions
A2	Regionally oriented	Continuously increasing	Slower and more fragmented than A1, B1, and B2	Self-reliance and preservation of local identities
B1	Rapid change toward service and information economy	Same as A1	Introduction of clean and resource-efficient technologies	Global solutions to economic, social, and environmental sustainability
B2	Intermediate levels of economic development	Continuously increasing, but not as fast as A2	Less rapid and more diverse changes than A1 and B1	Local solutions to economic, social, and environmental sustainability

Source: Intergovernmental Panel on Climate Change (IPCC), *Special Report on Emissions Scenarios (SRES)*, eds. Nebojsa Nakicenovic and Rob Swart (Cambridge: Cambridge University Press, 2000), http://www.ipcc.ch/ipccreports/sres/emission

Including sulfate aerosols in the models dampens the regional climate sensitivity, but greenhouse warming still dominates the changes. Models that include emissions of short-lived radiatively active gases and particles suggest that future climate changes could significantly increase maximum ozone levels in already polluted regions. Projected growth of emissions of radiatively active gases and particles in the models suggest that they may significantly influence the climate, even to 2100.

Stabilization emissions scenarios assume future emissions based on an internally consistent set of assumptions about driving forces (such as population, socioeconomic development, and technological change) and their key relationships. These emissions are constrained so that the resulting atmospheric concentrations of the substance level off at a predetermined value in the future. For example, if one assumes global CO_2 concentrations are stabilized at 450 parts per million (ppm) (the current value is about 380 ppm), the climate models can be tuned to produce this result. The tuned model predictions for regional climate changes can be used to assess specific impacts at this stabilization level. A more detailed discussion of the ability of the models to project regional climate changes can be found in Annex A.

Climate Projections of Future Temperature and Precipitation

The most recent IPCC report[xiii] states that the small islands of the Caribbean will probably experience a warming over the next century that may be somewhat smaller than the global annual mean warming. Temperature increases in the Caribbean at the end of the 21st century are projected to range from 1.4°C to 3.2°C with a median of 2.0°C. This level of warming is still likely to lead to significant sea level rise, deterioration of coastal areas, erosion of beaches, and increased invasion of non-native species. Reduced water resources could lead to an inadequacy of fresh water to meet demand during low-rainfall periods. The amount of sea level rise is not expected to be uniform because of the geographical differences in the islands. Extensive geographical, topographical, ecological, sociological, and population density information gathered into a detailed geographic information system (GIS) would be required before any predictions could be made.

Figure 12a shows the monthly changes projected for temperature and precipitation on a monthly basis from 1980-1999 to 2080-2099 in the Caribbean as reported by the IPCC.[xiv] Temperatures appear to change very little by month, unlike changes in precipitation. Most models predict changes in annual precipitation varying from –39 to +11 percent, with a median of –12 percent.[xv] Some regions are projected to have a slight increase in precipitation in December, January, and February (Figures 13b and 14), while a decrease is projected in June, July, and August.

Projections of decreasing precipitation in Central America and the Caribbean agree with projections of a general drying in the subtropical Atlantic associated with a phase shift to the positive phase of the North

Atlantic Oscillation (NAO). Increases in sea surface temperatures (SSTs) are of primary concern because of the relationship of SSTs to storm intensities. A projected climatological analysis of the Caribbean from 2041 to 2058 using a Parallel Climate Model (PCM) and National Center for Environmental Prediction (NCEP) reanalysis data showed a future warming of around 1°C (SSTs) along with an increase in the rain production during the Caribbean wet seasons. Although the PCM appears to under-predict SSTs, projected changes in feedback processes of cloud formation and solar radiative interactions lead to changes in projected rainfall variability and conditions that may be favorable for increases in tropical storm frequency.[xvi]

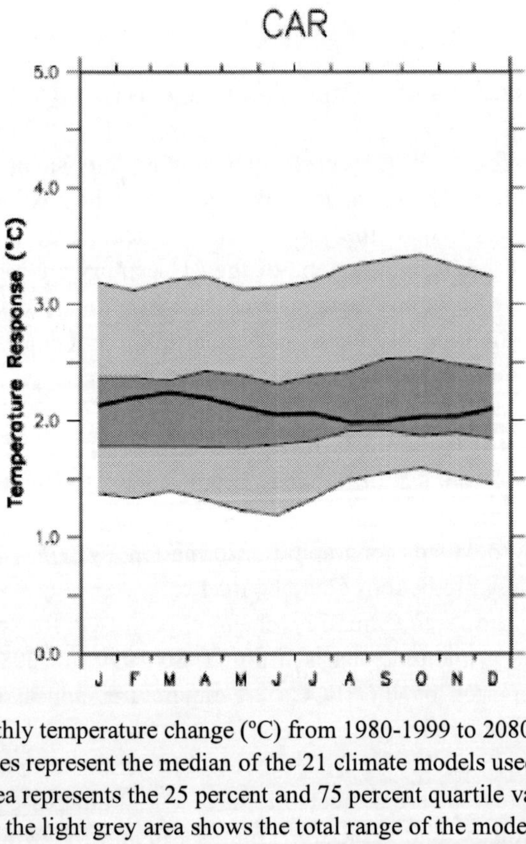

Figure 12a. Monthly temperature change (ºC) from 1980-1999 to 2080-2099 Caribbean (CAR). Thick lines represent the median of the 21 climate models used in the dataset. The dark grey area represents the 25 percent and 75 percent quartile values among the 21 models, while the light grey area shows the total range of the models. Source: J.C. and B. Hewitson, "Regional Climate Projections: Supplementary Material," in *Climate Change 2007: the Physical Science Basis*, eds. S. Solomon, D. Qin, M. Manning, M. Marquis, K. Averyt, M.M.B. Tignor, H.L. Miller Jr. and Z. Chen (Cambridge: Cambridge University Press 2007)

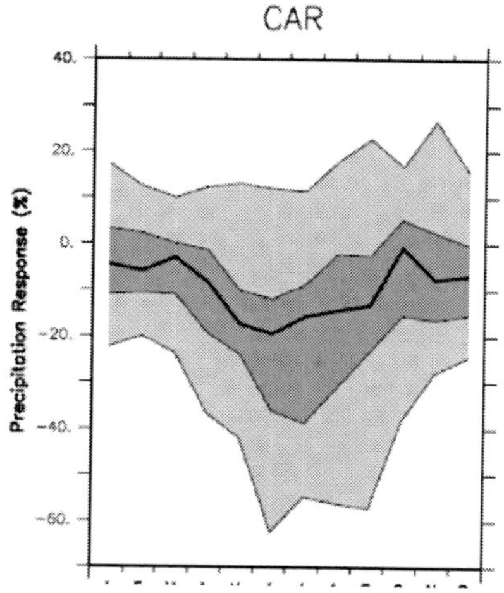

Figure 12b. As in 12a, but for precipitation change (%)

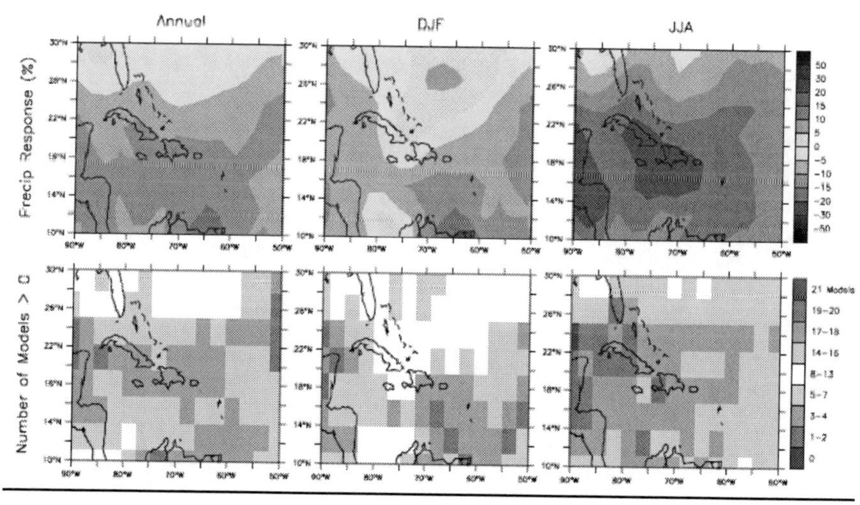

Figure 13. Projected precipitation changes over the Caribbean. Top row: Annual mean, December January February and June July August, fractional precipitation change between 1980 to 1999 and 2080 to 2099, averaged over the 21 climate models. Bottom row: number of models out of 21 that project increases in precipitation. Source: J.C. and B. Hewitson, "Regional Climate Projections," in *Climate Change 2007: the Physical Science Basis*, eds. S. Solomon, D. Qin, M. Manning, M. Marquis, K. Averyt, M.M.B. Tignor, H.L. Miller Jr. and Z. Chen (Cambridge: Cambridge University Press 2007)

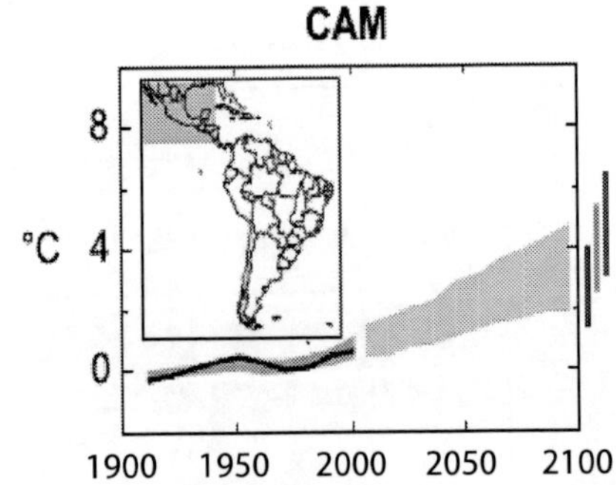

Figure 14. Temperature anomalies for Central America with respect to 1901 to 1950 for 1906 to 2005 (black line) and as simulated (red envelope) by models for 2001 to 2100. The bars at the end of the orange envelope represent the range of projected changes for 2091 to 2100 for various scenarios. Source: J.H. Christensen and B. Hewitson, "Regional Climate Projections," in *Climate Change 2007: the Physical Science Basis*, eds. S. Solomon, D. Qin, M. Manning, M. Marquis, K. Averyt, M.M.B. Tignor, H.L. Miller Jr. and Z. Chen (Cambridge: Cambridge University Press 2007)

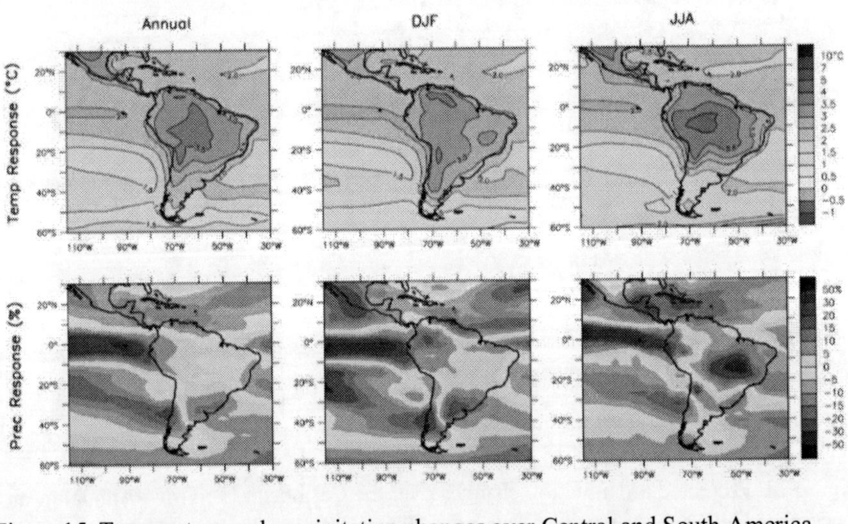

Figure 15. Temperature and precipitation changes over Central and South America. Top row: Annual mean, Dec, Jan Feb and Jun, Jul, Aug temperature change between 1980 to 1999 and 2080 to 2099, averaged over 21 models. Middle row: same as top, but for fractional change in precipitation. Source: J.C. and B. Hewitson, "Regional Climate Projections," in *Climate Change 2007: the Physical Science Basis*, eds. S. Solomon, D. Qin, M. Manning, M. Marquis, K. Averyt, M.M.B. Tignor, H.L. Miller Jr. and Z. Chen (Cambridge: Cambridge University Press 2007)

The IPCC projects a mean warming in Central America between 1980-1999 and 2080-2099 to vary from 1.8°C to 5.0°C, with half of the models projecting a range of 2.6°C to 3.6°C and a median of 3.2°C (Figure 14). There is a seasonal difference of around 1°C in the median values between winter (December, January, and February) and spring (March, April, and May). As projected for the Caribbean, Central America is likely to experience a decrease in rainfall in the future. Precipitation changes for Central America are shown in Figure 15.

The UK Hadley Centre PRECIS (Providing Regional Climates for Impact Studies) regional model was used to study climate change in Central America.[xvii] The researchers found that interactions between regional atmospheric circulation patterns, trade winds, and the region's complex topography not only define different precipitation regimes for the Caribbean basin (windward) and the Pacific basin (leeward), but also modify the annual cycle of precipitation. Assuming a doubled CO_2 environment, preliminary findings revealed that precipitation change in the future is very different on the Atlantic and Pacific sides of Central America and is also a function of elevation. The Atlantic side not only experiences a reduction in precipitation throughout the year, but also sees a change in the shape of the annual cycle where the Mid- Summer Drought feature seems to disappear. High elevation regions were shown to have an even greater reduction in precipitation compared to lowlands. This variability in the rainy season is very important for planning in key sectors, such as agriculture and power generation that are at the heart of the region's economy.

The same model was applied to Costa Rica[xviii] where cloud formation at high elevations is a primary source of moisture. Research indicates rising temperatures can cause clouds to form at higher altitudes, having a drying effect on areas below. These changes are expected to degrade the viability of numerous biological species in the area.

In 2003, the Hadley Center ran its climate change model using the scenarios from the IPCC Special Report on Emissions Scenarios (SRES).[xix] The Center concluded that the nominal warming predicted for all scenarios is similar over the next 40 years, even though each scenario represents a significant difference in level of emissions. This is explained by the long life of atmospheric CO_2 and the inertia of the climate system from emissions at the time of the study. The Center also concluded that the climate outcome for the second part of the 21st century will depend on the level of emissions in the next few decades. The model predicts precipitation changes in Central America and the Caribbean of up to -24 mm between present day and 2080s for the SRES A1B scenario. The important message from this modeling and analysis is that there is significant need to track, monitor, and mitigate the effects of rising temperatures and climate change at a country-by-country level.

Projections of Changes in Agricultural Growing Seasons

Central America is likely to continue converting forests for agricultural use. However, the general projected drying trend in the area is likely to limit the agricultural crops that can be grown. Projected temperature changes may not differ much by season, but changes in rainfall likely will. The result will be extended periods of drought and possible loss of soil fertility during the peak growing season in June, July, and August.

Also threatening agricultural productivity is the possible salinization of ground water supplies due to climate change and sea-level rise.

Many Central American and Caribbean countries have major fishing industries. Climate change is likely to lead to changes in migration patterns and depth of fish stocks thereby hurting the fishing industry.

Changes in the Frequency or Strength of Extreme Climatic Events

While increasing sea surface temperatures are linked to increasing storm intensities, natural variability in the coupled ocean-atmosphere system also plays a major role in hurricane variability. However, even considering the influence of natural variability, there has been a significant increase in Atlantic hurricane activity since 1970.[xx] During the 2005 hurricane season, SSTs across the tropical Atlantic were 0.9°C above the 1901-1970 average.[xxi] A recent study attempted to separate out the fraction of SST increase due to greenhouse-gas-driven climate change from that due to natural variations.[xxii] Results suggested that 0.45°C of the temperature increase in SST was due to global warming; El Niño accounted for about 0.2°C; the Atlantic Multidecadal Oscillation (AMO), explained less than 0.1°C, and year-to-year variability in temperatures explained the rest. This study contends that hurricane seasons will become more active as global temperatures rise. At the same time, however, there is still a great deal of debate in the scientific community regarding recent and future trends in hurricane frequency and intensity.

IMPACT BY COUNTRY OF CLIMATE CHANGE ON HUMAN-NATURAL SYSTEMS

This section examines the impact country-by-country, relying principally on insights provided in the submissions (National Communications) of the countries to the IPCC. Text boxes are included to highlight case studies and to include Puerto Rico in the discussion.

The submissions of countries to the UNFCCC provide national-level analyses driven by climate change scenarios. These submissions represent both high-quality scientific research and a degree of comparability not available in more local-level studies, which are few with the exception of Mexico.

Most of the impacts and vulnerability studies reviewed here use the IS92 scenarios.[xxiii] These scenarios (six alternatives, IS92a-f) were published in the 1992 Supplementary Report to the IPCC Assessment. The scenarios showed the evolution of greenhouse gas emissions over time, given assumptions about population and affluence. All of them assumed that no special policies to respond to climate change had been adopted. The resulting range of possible greenhouse gas futures spans almost an order of magnitude. Data came mostly from the published forecasts of major international organizations or from published expert analyses. IS92a has been widely used in impact assessments and assumes global population rises to 11.3 billion by 2100 and annual economic growth averages 2.3 percent between 1990 and 2100. Both conventional and renewable energy sources are used. The IS92e scenario has the highest greenhouse gas emissions, with moderate population growth, high economic growth, high fossil fuel availability and eventual phase-out of nuclear power. The IS92c scenario, on the other hand, has CO_2 emissions eventually falling below their 1990 starting level, with population first growing and then declining, low economic growth, and severe constraints on fossil fuel supply.

Belize

The Government of Belize completed its First National Communication to the Conference of the Parties of the UNFCCC[xxiv] in July 2002. The overarching conclusion of this first assessment is that the country's economy is highly dependent on a stable climate for successful agriculture, fishery, timber, and tourism industries. More than 50 percent of the country's GDP comes from the services industries where tourism plays a critical role. The country considers this assessment to be an initial effort to understand the role that Belize plays in the generation of greenhouse gas emissions, its adaptive capacity, the impact climate change variability will have on all economic sectors, and its human development goals. The government concedes much work is needed to understand the full impact and adaptability options.

The Minister of Natural Resources, Environment, Commerce and Industry stated the following: "Belize is prepared to continue working with the international community to negotiate responsibly for strong, achievable and enforceable mechanisms that will control the emissions of greenhouse gases. We are also prepared to utilize the nation's natural resources to assist in the global effort to mitigate the emissions as long as the measures can be accommodated within the nation's development strategy and ultimately contribute to the socio-economic development of our people."

The initial assessment was bounded by several key characteristics of the country: about 70 percent is still under natural vegetation cover, it has extensive low-lying coastal areas, and about 50 percent of its total population lives in urban centers along the coastal areas.

Belize is a net remover of greenhouse gases. In 1994, it was estimated that it absorbed six million metric tons (MMT) against three MMT of emissions. The Global Warming Potential (GWP), however, reveals a different picture. The GWP is a factor based on the relative radiative force for each gas and its respective life in the atmosphere. Using the GWP, Belize contributes to 9.5 MMT CO_2 equivalent while absorbing 3.5 MMT. Moreover, the UNFCCC recognizes that countries such as Belize, "Non-Annex I Parties," have a higher commitment to the alleviation of poverty and investing in sustainable development than to the mitigation of greenhouse gases.

Since signing and ratifying the UNFCCC, Belize has undertaken impact assessments on staple crops, coastal sensitivity to sea level rise, and water resources of the Belize River Valley. Climate change scenarios that project global mean surface temperature increases of 1°C to 3.5°C by 2100 are expected to contribute to a rise in sea level between 20 and 100 cm. Rising sea levels will have large effects on Belize's already low-lying coastline and its small islands with fragile ecosystems. Today, about 60 percent of the coastal areas experience flooding. Most residential areas around Belize City are built on drained/reclaimed wetlands vulnerable to sea level rise. A 1-meter rise in sea level would turn the wetlands into lakes, accelerating coastal erosion, exacerbating coastal flooding, raising water tables, and increasing the salinity of rivers and aquifers. This rise would also provide a higher level for coastal flooding, forcing storm surges further inland and facilitating greater damage from smaller surges.

In the past 20 years, Belize's rate of real estate development (hotels, restaurants, tourism services) in the coastal areas has accelerated sharply to accommodate the growing tourism industry and the expansion of coastal residential areas. The rapid growth has placed increased pressure on the available resources manifested by reduction in water quality, increased soil erosion, and an overextended waste disposal infrastructure. The study estimates that a 50 cm rise in sea level over the next 100 years would overtake more than 50 percent of the beaches; a 100 cm rise would destroy 90 percent of the beaches.

The outlying islands and the Placentia Peninsula are already threatened by a 20 cm rise in sea level. The conclusion at this time is that, to protect these urban areas, sea walls and dikes will need to be built.

When the simulation model that was used adds rising sea levels to increased precipitation as expected in rising temperatures, it reveals that the river areas of the country will remain in a permanent state of flooding throughout the year because of reduced drainage capacity.

Saltwater intrusion is another major concern throughout the coastal areas. Some of the outlying islands have already been equipped with desalination plants to reduce the impact of growing demand on drinkable water by development/population expansion. At this stage, it is clear that the projected sea level rise in the next 100 years, coupled with increases in the rate of water extraction, will result in higher events of saltwater intrusion. Belize gets its water upstream where the water is already salty during the dry season, making drinking water salinity a problem.

Aquaculture has been undertaken along the coastline in areas that are vulnerable to flooding and erosion. Together, these increase water turbidity, which in turn reduces the productivity of cage aquaculture and fish/shrimp farms along the coasts.

Belize's coral reefs are not expected to suffer from rising sea level, but from rising temperatures and rising storm surges. Its coral reefs are living near or at their upper temperature resilience today, so a small increase in temperature will cause them to "bleach," making the corals more susceptible to diseases/pathogens that would eventually kill them. Two bleaching events occurred in Belize in 1995 and 1998 (ENSO years), and elevated sea temperatures affected 52 percent of the reefs. The economic impact of losing coral reefs is twofold: aquaculture and tourism. Tourism today accounts for 15 percent of the GDP and is the largest source of foreign exchange and employment.

Cuba

The Government of Cuba submitted its First National Communication to the UNFCCC[xxv] in August of 2004. The study included the main island and all adjacent islands that form the Cuban Archipelago. Cuba's climate is tropical, with marine influence and average temperatures ranging from 24°C in the plains to 26°C and slightly higher in the eastern shores. The variability in climate stems mostly from the level of precipitation. The average annual rainfall is 130 cm/year between May and October, when 80 percent of total precipitation occurs. The dry period runs between November and April.

The most common and frequent weather events that occur in Cuba are tropical cyclones. This is the term used for the different levels of intensity from tropical depression to hurricane. From year to year, Cuba can experience 0-5 tropical depressions/storms and 0-4 hurricanes.

The Cuban economy suffered a severe contraction with the breakup of the Soviet Union. The contraction triggered a large reduction in funds and goods injection, eliminated the trading links to eastern European countries, and limited access to external credit. All of this caused Cuba's GDP to experience a freefall between 1989 and 1994. In 1995 the economy started to recover slowly and has continued with small but positive GDP changes.

In its National Communication, the IPCC main categories of greenhouse gases were used to calculate national inventories with the following activities as sources: energy, industrial process, solvents and other product use, agriculture, land-use change and forestry, and waste.

Greenhouse gases were estimated at 41,314 gigagrams (Gg) in 1990 compared to 26,043 Gg in 1994. CO_2 was the greatest contributor to emissions (94 percent) from the energy sector in both years, though a net removal of gases was achieved by the changes in land use and forestry sectors. The 37 percent decrease in that timeframe resulted from the sharp economic contraction.

Initial estimates of future greenhouse gases, with annual GDP growth of 4-6 percent and carbon intensity levels equal to the ones in 1990 and no mitigation efforts, indicate that Cuba's gross level of emissions will reach 81.3 MMT by the year 2020. When the simulation model includes a reduction on the real energy intensity achieved since 1990, the gross emissions levels drop to around 70 MMT. This implies that there is potential for greenhouse gas reduction by the year 2030.

Cuba enjoys a robust network of surveillance systems focused on meteorology, climate, and atmospheric pollution with 75 meteorological stations and 11 rain and air quality monitoring stations. This surveillance network accounts for a significant contribution of information and data to the World Meteorological Surveillance System (WMS), the Global Atmosphere Surveillance (GAS), the Global Climate Observing Systems (GCOS) and the Global Ocean Observing System (GOOS).

Cuba has a well-structured system of research programs that covers a wide variety of problems focused on understanding the economic, technical, intellectual, and cultural development of the country. The following studies are underway:

- Global change and the evolution of the Cuban environment.
- Sustainable development of the mountains.
- Sustainable energy development.
- Production of foods for its population through sustainable ways.
- Production of animal food through sustainable ways.
- Agricultural biotechnology.
- Vegetable improvements and phytogenetic resources.
- Biodiversity.
- Agricultural ecosystems and soils.

Cuba conducted a study on national biodiversity in 1995 and learned that it has 6,700 plant species, 42 different ecosystems, and more than 19,600 animal species. About 10 percent of the animal species and 2 percent of plant species are at risk of extinction. Cuba classifies 30.8 percent of its agricultural land as “low productivity” and 46 percent as “very low productivity.” This has resulted in a continued crop productivity loss with yield indexes below 70 percent.

Observations confirm that Cuba has experienced an annual average temperature increase of about 0.5°C during the period 1951-1996. This is attributable mainly to an increase of 1.4°C in average minimum temperatures, while the increase in average maximum temperature has been insignificant, i.e., there is a reduction in the daily variability of temperatures. Cuba at the same time has experienced an increase in the severity of events such as tornados, rain, hail, and drought since the mid-1070s. ENSO played a key role in the climate variability across the country during this period.

The study, initiated in 2000 to simulate future effects of climate change, used the MAGICC/SCENGEN climate models to generate three different scenarios—optimistic, moderate, and pessimistic. The scenarios combined increases in temperature, using 1990 as the base year, for the years 2010, 2030, 2050, and 2100, and their corresponding rise in sea level for both IS92a and KyotoA1 emissions scenarios. The variance in temperature increase from the three scenarios ranged from 0.34°C to 2.52°C. The variance in sea level rise from the three scenarios ranged from 2 to 55 centimeters. Since all models have limitations, the above scenarios were evaluated with two separate models.

Climate change in general tends to decrease the amount of surface water, even in the case where the model projects precipitation increases. Saltwater intrusion into aquifers is a serious concern and highly probable because most of Cuba's aquifers are open to the sea. A rise in sea level of 30 cm by 2100 will result in a rise in saltwater intrusion of no less than 10 miles inland.

The impact on coastal zones and marine resources based on the scenarios evaluated can be summarized as flooding and displacement in low-lying areas, coastal erosion and the retreat of the coast line, an increase in storm surges, an increase in the salinity in estuaries and aquifers, changes in sediment patterns, and the reduction of light in the marine ecosystem.

The study evaluated the impact in agriculture by focusing on food crop productivity, biomass, diseases and pathogens, and forests. For food production, a set of basic products such as beans, soybeans, corn, cassava, sugar cane, rice, potatoes, and sorghum was evaluated. Productivity losses for the year 2030 where there was no fertilizing effect from CO_2, was between 10-15 percent for rice, cassava and corn; 5-10 percent for sugar cane; and 40-45 percent for potatoes. If the model includes the fertilizing effect from CO_2 and crops, such as beans, soybeans and rice with shorter growing cycles, gains in productivity are possible. These results will depend on the sensitivity of the climate to changes in energy balance.

Mangroves and other forest areas (especially deciduous trees) will suffer. The National Communication estimates that by 2030 sea level rise will affect 7.1 percent of mangrove forest areas, with about 42.9 percent of the area not recoverable.

The evaluation of the impact of climate change on human health focused on the following diseases: acute lung diseases, bronchitis, viral hepatitis, chicken pox, meningitis, and acute diarrhea. Initial estimates revealed that all of the diseases would almost double by 2010 from the 1991-1998 base year with implications of similar impact on the cost of dealing with the increased number of disease events.

Dominican Republic

The Government of the Dominican Republic completed its First National Communication to the UNFCCC[xxvi] in March 2004. The Dominican Republic is an island nation whose variable climate is highly influenced by the surrounding water, easterly winds, pressure systems, topography, and recurrent hurricanes. Its average annual precipitation is 150 cm that varies from 35 cm to 274.3 cm (~ 108 in) in the island's interior mountain range. Emissions for 1990 and 1994 were estimated at 8,690 Gg and 15,003 Gg, respectively. More than 90 percent of the emissions were of CO_2, mostly from the use of fossil fuels to meet energy demands. Similar to the rest of the countries in the region, total gross emissions are very small, but the effects of climate variability can be significant.

The study to evaluate the impact of climate change was performed using different scenarios with adjustments for the Dominican Republic's climate patterns. The study evaluated the effects on water resources, coastal zones, agriculture, forestry, and health.

Three emission scenarios were chosen for this assessment, an optimistic one, IS92c; a moderate one, IS92a; and a pessimistic one, IS92f. The base period chosen was 1961-1991. Projections were made for temperature, precipitation, and rise in sea level. Under the moderate scenario temperature is expected to increase to 26.9°C, precipitation to decrease to 113.7 cm, and sea level to rise by 12.33 cm by 2030. These projections are compared to actual levels for 1990.

The evaluation on water resources used a methodology that included the current water balance adjusted with coefficients representing average monthly changes in temperature and rainfall. Three models were used—CSRT, ECH4 and HADCM2—for the IS92 emissions scenario (moderate) at different levels of sensitivity. The models were run against two regions of the country. The models were run for the years 2010, 2030, 2050, and 2100. A rise in sea level similar to the one used and observed in Cuba (2.9 mm/year) was used to evaluate the impact on aquifers. Rainfall is the only source of water replenishment in the Dominican Republic. Since moderate scenarios estimate a reduction of up to 25 percent in water resources, the Dominican Republic will need new policies to reduce water demand and will have to invest in infrastructure to increase its supply.

The most important aquifers in the country are open, which means they are in contact with ocean water. This is why saltwater intrusion will increase with rising sea levels, exacerbating the loss of water resources for urban, industrial, and agricultural use. The CSRT model estimates an increase in temperature of 0.7°C and a 4 percent increase in rainfall. The ECH4 model estimates an increase in temperature of 2.6°C and a 10 percent increase in rainfall in the next 100 years. As temperatures rise, so does evaporation, reducing water resources by 28 percent relative to the base period. The HADCM2 model estimates a rise in average temperature of 4.2°C and a decrease in rainfall of 60 percent, causing a loss of 95 percent of the water resources by the year 2100.

The Dominican Republic's coastal areas have a rich diversity in ecosystems and economic activities. The cities and towns located along the coastal areas contain 64 percent of the population. Based on the chosen scenarios of emissions and rising sea levels of 0.14 cm/year and 1.01 cm/year, a cumulative rise in sea level is projected to range from 3.77 cm to 26.73 cm by the year 2030. This will affect major coastal roads, housing, and bridges with all needing repairs and reconstruction more frequently.

The Dominican Republic enjoys an active and growing tourism industry. The majority of the activities are associated with beaches, coral reefs and clear water. Lack of information on soil erosion and coastal erosion did not allow for a complete evaluation of the potential impact of rising temperatures and rising sea levels on the tourism economy. At this time, rising temperatures and rising sea levels are not expected to have a large impact on the fishing industry.

Forest productivity in the Dominican Republic today is very high in areas of large rainfall and very low in areas of low rainfall. Under the scenario used for the HADCM2 model (including the fertilizing effects of CO_2), estimates of up to 21.2 percent increases in forest productivity are projected by the year 2050 in the regions where there is currently a high level of production. There is no significant change in productivity in the regions of low yields.

Evaluation of the impact on agricultural production was focused on potatoes, rice, and corn as initial examples since the current methodology could adjust for the variability across the country on growing cycles, rainfall, and products. The impact on potatoes under all scenarios is negative. The largest decrease is associated with the HADCM2 model that projects that in the latter part of the next century growing potatoes may be impossible. Productivity losses by the year 2030 are estimated at above 50 percent. Productivity losses for rice are less dramatic, ranging from 12 percent by the year 2030 to around 50 percent by the year 2100. There are no significant productivity losses for corn; however, productivity does gradually decrease in the same timeframe.

The impact on health in the Dominican Republic was focused on the patterns observed in the past 10 to 15 years. In some regions, 80 percent of malaria cases have been observed where only 10 to 15 percent of the population is located. Changes in temperature and rainfall as projected by the scenarios reveal that future adjustments would be in the frequency of cases (increasing from 16 to 20 percent), but that current patterns and geographical distribution would be maintained.

In this analysis, there was no clear evidence or correlation between ENSO and increases in the number of malaria cases.

Guatemala

Guatemala submitted its first national communication to the UNFCC in December 2001.[xxvii] The communication, prepared by the Ministry of Environment and Natural Resources, identifies four 41 major areas vulnerable to climate change: health, forests, production of basic grains, and hydrologic resources. The analysis of Guatemala's climate is based on data from the network of stations of the National Institute of Seismology, Volcanology, Meteorology and Hydrology (Instituto Nacional de Sismología, Vulcanología, Meteorología e Hidrología). The emissions scenarios chosen to estimate the changes in global mean temperature were IS92c, IS92a, and IS92d.

Three diseases among several analyzed were identified as the principal diseases associated with climatic variability in Guatemala: acute diarrhea, acute respiratory infection, and malaria. Acute diarrhea and malaria are highly prevalent in the warmest and rainiest months of the year (MayOctober).[xxviii] Acute respiratory infection is more prevalent during September-November and February-March (i.e., the transitioning periods from summer to winter and vice versa). The criteria used to assess the diseases with the largest possible impact from climate change are the following:

- Having a relation (direct or indirect) to climate and its variability.
- Being of high prevalence at a national level.
- Being within the ten major causes of morbidity and mortality.
- Having statistical data to develop the research.
- Not having been discarded in other studies at the international level.
- Having a profound understanding of its epidemiology.
- Obtaining results that are beneficial to health.

In addition to coinciding with the largest number of the aforementioned criteria, acute diarrhea and acute respiratory infection represent the studied diseases with the greatest potential impact on the country and are the principal causes of illness and deaths among Hondurans (363,679 and 962,827 deaths in 1999, respectively), particularly in children less than 5 years old. Malaria was selected because it is a vector disease and is predominantly present among adult males. The poor state of health in Guatemala results from the poor quality of life, little availability of health and sanitary programs, and the lack of adequate nutrition. The prevalence of infectious diseases in Guatemala reflects this poor state of health and is among the major causes of deaths in the country.

An assessment of the impact on health due to climate change was based on climatology baseline from 1961-1990, climate information for the period 1991-1999, and a pessimistic scenario on climate change. The Bultó Index was also used. This is a methodology developed in Cuba that is based on empirical statistical models for projecting future behaviors of diseases using climatic conditions as variables, such as maximum and minimum temperatures, thermal oscillation, precipitation, and the influence of ENSO. The analysis of the three diseases focused on the southwestern part of Guatemala, but according to the communication, the results could be similar for the rest of the country with the exception of malaria which shows a tendency to decline by 2030.

The results indicate that by 2030, acute respiratory infection has a tendency to increase in frequency, particularly at the beginning of the rainy season. By the same year, acute diarrhea also increases in frequency, with the greatest prevalence in June and July. However, the communication points out that the disease is not only influenced by climate and its variability, but also by such factors as poor sanitary infrastructure, especially in rural and marginal urban areas of the country, the lack of health education, and poor coverage of health care services. When analyzing the effects of climate variability in the behavior of malaria in the region, the result is a significant decline in malaria cases and abating of the seasonal patterns of the disease as a result of the effects of climatic variability.

For the assessment of climate change impacts on forest resources, vegetation cover is analyzed as a function of the IS92c, IS92a, and IS92d climate change scenarios and bioclimatic scenarios assigned to Guatemala that are based on the Holdridge Life Zone Model. Under the optimistic scenario, climatic conditions have an impact on very limited areas of the country; only 416 km^2 of forest cover (0.38 percent of the total surface area of the country), which is equivalent to 4.2 million cubic meters of lumber. Under the pessimistic scenario, close to 4,000 km^2 of coniferous and mixed forests (3.67 percent of the surface area of the country) would suffer, which is equivalent to 40 million cubic meters of lumber. Coniferous forests represent 80 percent of forest productivity, so the decline in forest cover would also have economic consequences. However, the authors caution that the analysis is based on climatic projections to 50 years (from 1999), which is a short period of time for forests to show significant changes.

The climate change scenarios for the vulnerability study on the production of basic grains is based on the changes for the year 2030 in the normal (ECCG_C), optimistic (ECCG_HA, extensive wetness), and pessimistic (ECCG_SA, extensive dryness) scenarios.[6] Corn, beans, and rice, with their cultural, socioeconomic, and nutritional significance in Guatemala, were the basic crops studied. Corn is the most important crop in the country and makes up the basic diet of a majority of Guatemalans, particularly in rural areas. In addition, most corn producers are subsistence farmers. Beans are the second most important food crop in Guatemala and one of the major sources of protein. Rice is a significant source of carbohydrates in the national diet and is also used in the production of domestic beverages.

The yield differences in the production of basic crops that were simulated according to the baseline (projections of environmental conditions in the absence of climate change), and the differences that were obtained under climate change, represent the potential size of the impact for 2030 (Table 6).

For each zone examined, yield variability (production) was determined between the production in the baseline and production under climate change, and for the normal, optimistic and pessimistic scenarios (Table 7).

The simulations show that variability in normal climate conditions implies the largest negative consequences on the studied crops. The results also indicate that in the areas where climatic conditions are expected to be more extreme, the extent of negative consequences on agricultural production of basic crops will be larger.

The hydrometeorological data used to evaluate the impact of climate change on hydrological resources are based on the results (precipitation and evapotranspiration) of the baseline of the climate scenarios, which was used to create a base scenario for precipitation (P), evapotranspiration (ETP) and runoff (R) and for each basin studied. The MOD-BAL model, developed by

[6] The scenarios were defined specifically for the National Communication; details can be found in that report.

UNESCO, was used to estimate future runoff according to climatic parameters established in the climatic scenarios to the year 2030.

Table 6. The simulations were made for corn, beans and rice, for13 agricultural seasons (1980 to 1993), and in seven climatic observatory sites

Station	Temperature increments (ºC)			Precipitation Variability (%)		
	HA	C	SA	HA	C	SA
Camantulul	1.5	1.0	2.2	+9	-1	-19
Panzós	1.6	1.0	2.1	-1	-2	-19
Asunción Mita	1.6	0.9	2.3	+9	-2	-22
Labor Ovalle	2.8	2.4	3.6	+7	-1	-19
San Jerónimo	1.4	1.1	2.3	+6	0	-10
INSIVUMEH	1.5	1.0	2.2	+7	-1	-18
S. Cruz Balanyá	1.5	1.0	2.2	+7	-1	-18
Promedio	1.7	1.2	2.4	+6	-1	-18

HA: Optimistic scenario (excess wetness); C: Normal scenario (central); SA: Pesimistic scenario (excess dyness)

Source: Herrera and Associates (2000)

Table 7. Climate change impacts on the production of Basic Grains

Zone	Crop	Yield (Kg/ha)							
		Actual	Baseline	Optimistic	% Change	Normal	% Change	Pessimistic	% Change
1	Corn	2857	2738	3142	15	2957	8	3091	13
2	Corn	2025	1952	1744	-11	1828	-6	1630	-16
	Rice	2025	4136	3303	-20	3462	-16	3018	-27
3	Corn	2270	2263	2029	-10	2003	-11	1500	-34
	Beans	1281	1281	743	-42	918	-28	433	-66
4	Corn	2189	2163	2430	12	2280	5	2131	-1
5	Corn	1954	1954	2021	3	1918	-2	1876	-4
6	Corn	2237	2245	2156	-4	2169	-3	2120	-6
7	Corn	2384	2374	2412	2	2447	3	2339	-1
6	Beans	113	2104	2157	3	2163	3	2110	00

Source: Herrera and Associates (2000).

Under the optimistic scenario (ECCG_HA), an increase in runoff can be expected. River flows of 10 liters per second under this scenario would increase to up to 11.5 liters per second. Under the pessimistic scenario (ECCG_SA), a reduction in runoff can be expected. Runoff of major rivers of large departments and cities such as Guatemala, Escuintla, Mazatenango and Quetzaltenango may decrease by as much as 50 percent. Accordingly, basins of 10 liters per second could diminish by as much as 5 liters per second.

Haiti

Haiti's Ministry of the Environment submitted the country's First National Communication to the UNFCCC in 2001.[xxix] Haiti has also submitted a National Adaptation Programme of Action (NAPA) to the UNFCCC.[xxx] NAPAs provide a process for Least Developed Countries (LDCs) to identify priority activities that respond to their urgent and immediate needs with regard to adaptation to climate change. The rationale for NAPAs rests on the limited ability of LDCs to adapt to the adverse effects of climate change. In Central America and the Caribbean, only Haiti has been designated as an LDC.[xxxi]

Similar to many of the Latin American reports, Haiti's National Communication first focuses on emissions sources, particularly energy emissions. Four sources are listed: wood (71 percent), oil (20 percent), hydropower (5 percent), and bagasse (4 percent). Most of the energy demand (69 percent) is for residential housing.

The vulnerability section focuses on agriculture and water. The method used is also the method specified for many of the other countries in this region, using the MAGICC and SCENGEN models to generate scenarios. There are three sensitivity scenarios (low, middle, and high). The results show temperature increases of ~0.6 to 1.2°C by 2020, ~1.1 to 2.3°C by 2050, and ~1.4 to 4.0° by 2100. By 2030, precipitation at a medium sensitivity decreases from 5.9 percent in February to 20 percent in July.

For agriculture, all three crops studied—potatoes, rice, and maize—show decreased yields, even with CO_2 fertilization. Forestry, too, is projected to experience detrimental effects. Less precipitation and higher temperatures are the sources of these negative consequences.

For water, the decline in precipitation has a devastating impact, combined with saltwater intrusion as sea level rises. Every variable shows marked changes; for instance, precipitation in 2030 is projected to decline by 187 mm annually and continue to decline to 477 mm annually by 2060.

Honduras

Honduras submitted its first national communication to the UNFCC on November 15, 2000.[xxxii] In considering possible effects of climate change, the communication draws partly on projections developed by the country's Climate Change Program of the Environment Ministry and partly on a 1995 US EPA-funded Central American Project on Climate Change (Proyecto Centro Americano de Cambio Climático). Under the Central American Project on Climate Change, Honduras participated in studies on the vulnerabilities of hydrologic resources and addressed the possible impact from climate change-related sea level rise.

Honduras is highly affected by extreme climatic events—in terms of both the frequency of climatic changes, as well as the intensity of occurrences. In 1995-96, the impact from drought in the driest regions of the country brought about famine, human losses, emergence of water-borne diseases, cardiovascular and respiratory diseases related to atmospheric pollution and extreme temperatures, loss of crops, and increased forest fires. Hurricane Mitch in 1999 and the historic amounts of rainfall that followed the next year cost the lives of many civilians, as well as causing appreciable losses and deterioration of infrastructure, crop failure and depletion of watersheds.

In response to this, Honduras' Climate Change Program of the Environment Ministry has developed sectoral vulnerability studies associated with medium-to-long-term occurrences such as climate change. Studies specific to climate change relate to future climate projections based on the IPCC scenarios. The table below shows projected future changes of average annual temperature, precipitation, and cloudiness for the pessimistic and moderate scenarios based on these studies:

Pessimistic Scenario

Year	Temperature °C	Precipitation (%)	Cloudiness (%)
2010	0.6 to 0.9	-6.6 to -8.4	-2.5 to -4.0
2030	1.0 to 1.5	-11.2 to -14.5	-4.3 to -6.8

Moderate Scenario

Year	Temperature °C	Precipitation (%)	Cloudiness (%)
2010	0.6 to 0.8	-2.4 to -6.4	-2.4 to -3.7
2030	0.9 to 1.3	-9.7 to -12.5	-3.8 to -5.9

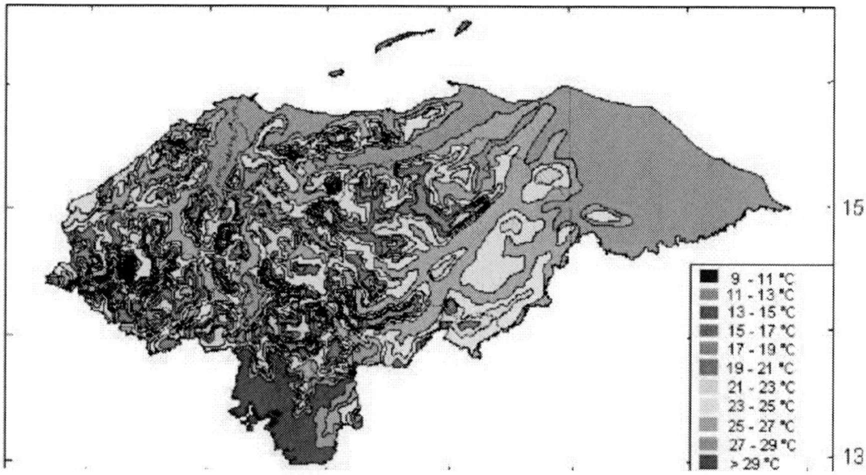

Source: Honduras, "First National Communication to the Conference of the Parties of the United Nations Framework Convention on Climate Change" (November 2000) http://unfccc.int/national _reports/non-annex _i_natcom/items/2979.php

Figure 16. Spatial distribution of temperature in Honduras – Results for the year 2030 given the moderate scenario

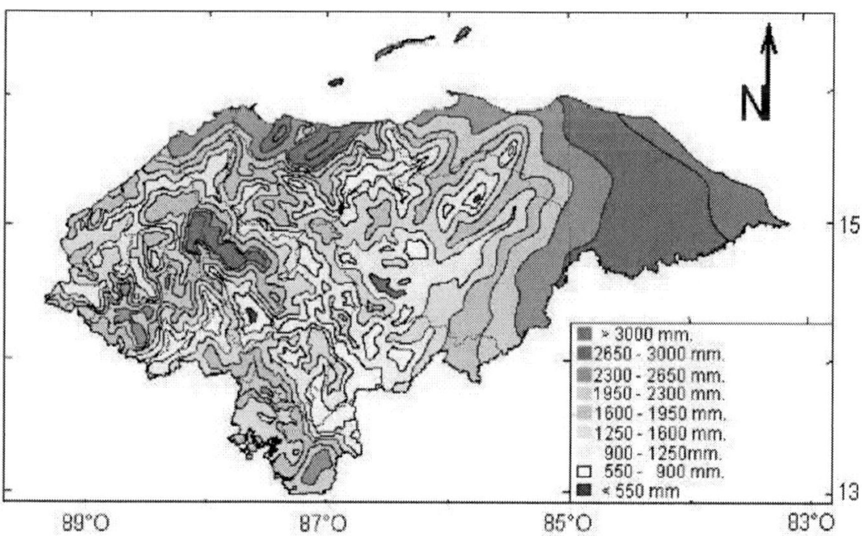

Source: Honduras, "First National Communication to the Conference of the Parties of the United Nations Framework Convention on Climate Change" (November 2000) http://unfccc.int/national _reports/non-annex _i_natcom/items/2979.php

Figure 17. Spatial distribution of precipitation in Honduras – Results for the year 2030 given the moderate scenario

Figures 18 and 19 identify what areas in Honduras are likely to experience the highest temperatures and the most precipitation for 2030, respectively.

Reduced precipitation as indicated by these projections may cause considerable sectoral damages, particularly if this reduction due to climate change is accompanied by precipitation reductions that arise from an El Niño event in areas near the Pacific slope.

Changes in the hydrologic cycle due to climate change will occur in the form of floods and droughts that year by year will affect considerably the agricultural zones of the country, such as the Valley of Comauagua, the Valley of Sula and the Valley of Choluteca. The rise in temperature and reduction in rainfall will likely have effects on the supply of water for drinking, irrigation, and the generation of electric energy. Given the high importance of agriculture to Honduras, it is highly likely the economy will suffer severely.

Based on the 1995 Central American Project on Climate Change, preliminary estimates of the areas vulnerable to sea level rise by 2025 are shown below.

Affected Zone	**Loss by inundation estimates (km^2)**
Valle de Sula	885
Valle de Cuyamel	39
Punta Gorda	3
Omoa	2
Tulián	3
Puerto Cortés	20
Bahía de Tela	46
Valle de Río Leán	100
Llanura del Esparta a la Ceiba	175
Total	**1,276**

In addition to the economic damage from flooding associated with the rise in sea level, Honduras' first communication also points out possible socio-cultural implications that are difficult to quantify. Such may be the case with nationally treasured archeological sites and valuable tourist resources, such as mangroves, wetlands and reefs that are vulnerable to erosion and flooding.

Mexico

The Government of Mexico submitted its Third Communication to UNFCC as an update to its previous submission[xxxiii] in December 2007. This report includes an inventory of greenhouse gases in 2002. In contrast with the other countries in this review, Mexico is a net producer/supplier of fossil fuels and an increasingly important emitter of greenhouse gases.

Inventories of greenhouse gas emissions for the third report were calculated for the year 2002 in the energy sector, industrial processes, solvents, agriculture, land use/changes/forestry and waste. The energy sector generated 61 percent of all emissions, followed by land use/changes/forestry with 14 percent, waste 10 percent, industrial processes 8 percent and agriculture 7 percent. At that time emissions in terms of CO_2 equivalent represented an increase of 25 percent from the base year (1990).

The Long-Range Energy Alternatives Planning (LEAP) system was used to build the base emission scenarios for 2008, 2010 and 2030 to estimate future greenhouse gases. For these projections three scenarios were used: base or current, low economic growth and high economic growth. A key conclusion was that electricity generation is highly sensitive to GDP growth, resulting in 30 percent reduction of emissions in the low economic growth scenario and 24 percent increase in the high economic scenario. Another conclusion accepted in the report found that implementing automobile energy efficiency standards would help significantly in the reduction of greenhouse gases, in combination with the expansion of renewable and nuclear energy. This report focused extensively in quantifying all types of gases generated by the use and generation of energy/fuels across the major sectors of the economy. Currently, there are many programs, regulations and measures underway by the Secretaría de Energía (SENER) to increase the efficient use of energy as well as to save energy with the goal of reducing greenhouse gas emissions in the next 100 years. There are also programs to increase the role of renewable energy including wind and biomass. By the end of 2007 the plan was to have a total of 5,000 GW/year for incorporation into the electricity network with the goal of covering 8 percent of total electricity demand by the year 2012.

GCMs were used to estimate the impact of climate change under four SRES emission scenarios: A1B, A2, B2 and B1. The overarching results were that Mexico's climate is projected to be warmer by the years 2020, 2030 and 2080, especially in the northern part of the country with a temperature increase of 2-4°C. Rainfall was estimated to decrease by 15 percent in the central regions and by less than 5 percent on the regions around the Gulf of Mexico. The hydrological cycle will be more intense creating a larger number of storms during the rainy season and a prolonged period of drought during the dry season. These cycles indicate that 75 percent of precipitation will evaporate while only 5 percent will be able to replenish aquifers. IPCC estimates that Mexico could experience reductions in runoff ranging from 10-20 percent as a national average with over 40 percent in coastal areas of the Gulf. The projected increases in severe storms and prolonged droughts made by these models have already been observed across the country in the past five years.

Mexico's national average water availability is calculated at 4,000 cubic meters per capita per year. The national average availability varies significantly among the different regions of the country, particularly in the center and north, where the average is 2,500 cubic meters per capita per year.[7] The figures are somewhat misleading since 75 percent of the water is used by the agricultural sector, 14 percent by households and 11 percent by the industrial sector. According to the Comisión Nacional del Agua (CNA) the agricultural sector wastes 55 percent of water extracted while the urban sector wastes 43 percent due to leaks in the extraction and distribution process and through excessive use. Adding decreases in rainfall of 5-10 percent and increases in temperature of 1-3ºC will result in water availability losses between 5 and 15 percent by the year 2020 and 2050. Water resource loss will vary widely across the country. This will have a critical impact in the north and central regions, a severe impact in the Pacific-central region, and a strong-to-moderate impact in the south and Gulf coast regions. As a result, it is clear that significant changes will be required in the use and distribution of this resource.

Climate change projections were applied to three models for evaluating the efficiency of corn yields based on temperature, rainfall, topography, soil type and vegetative period. Moderate yield losses were found in the moderate yield areas, which would force increased use of marginal lands of up to 4 percent, resulting in further yield reductions. Several models were used with the same scenarios, providing different results with variations of productivity changes between slight increase and moderate losses depending on the region.

Forestland coverage is a key factor in the mitigation of climate change. The risk of forest fires increases with rising temperature and reduction of rainfall. Loss of forestland will be exacerbated as agricultural activity moves into marginal lands and forest areas. Under the different models used for

[7] 1,000 cubic meters per capita per year is an indicator of water scarcity.

scenario A2 the estimated forestland affected ranges from 8 percent to 33 percent by 2020 and from 9 percent to 76 percent by 2050.

The report provides results of a specific study on water resources undertaken for the Hermosillo and Sonora regions. The study outlines options as solutions to the water availability challenges with some qualitative estimates of the type of action, time to implement, efficiency ratios, cost, viability, participants, and outcomes.

Other studies of Mexico generally echo the major concerns about impacts discussed in the National Communication, especially crop production, precipitation, and water availability. Conde et al.[xxxiv] focus on maize, the staple food of rural dwellers, especially subsistence farmers in Tlaxcala, Mexico. Mexican policy changed from self-sufficiency in food production during the 1990s to an emphasis on "guarantee[ing] people's capacity to acquire food."[xxxv] Imports became more important—but not to the poorest and subsistence farmers. Using the SRES A2 and B2 scenarios leads to projected yield increases because the threat of frost is reduced. Using the Ceres-maize model, however, leads to yield reductions. Wehbe et al.[xxxvi] explore coffee production in its climatic and economic context in Veracruz. Their model indicates that coffee production falls by 34 percent by 2020, making it not economically viable. Salinas-Zavala and Lluch-Cota[xxxvii] find that ENSO events are correlated to winter wheat yields in Sonora (El Niño with increases, La Niña with decreases); the ability to forecast ENSO events may thus reduce the impact of climate change on wheat yields. Luers et al.[xxxviii] also focus on wheat in the same region (the Yaqui Valley), specifying a quantitative measure of vulnerability and finding that "Valley farmers, without adaptations, are on average more vulnerable to a 20 percent decrease in wheat prices than a 1°C increase in average minimum temperature."[xxxix]

Drought has long plagued Mexico. Boyd and Ibarrarán[xl] explore the implications of projected increases in drought in northern Mexico (up to a 36 percent increase projected by the Canadian Climate Change model) on various economic sectors. As expected, agricultural production is highly affected. Electricity from hydropower constitutes another significant loss. A ripple effect then slows productivity in manufacturing, chemicals, and refining sectors, although these losses are not as great as in agriculture and electricity. Finally, consumption declines, with inequality increasing as the already-poor are more affected.

Nicaragua

A case study for evaluating impacts of climate change in Nicaragua[xli] revealed that temperature increases ranging from 1.3°C to 1.5°C by the year 2030 would result in a 12.4 percent to 14.5 percent drop in precipitation. In this study, Umaña and colleagues considered three main temperature change scenarios: optimist, moderate and pessimist for the years 2010, 2030, 2050 and 2100. The optimist scenario assumed temperature increases of 0.8°C for 2010, 1.3°C for 2030 and 2.1°C for 2100, resulting in 7.9, 12.4 and 21 percent decrease in precipitation.

For the moderate scenario with temperature changes of 0.8°C in 2010 and 1.3°C in 2030 (the same as for the optimist scenario), precipitation is estimated to decrease from 7.9 to 12.4 percent. For the pessimist scenario, temperature change is projected to increase by 0.9°C in 2010 and 1.5°C in 2030, resulting in a decrease from 8.4 percent to 14.5 percent in precipitation.

The scenario changes in temperature and precipitation were used to simulate the impact on the three main food crops produced in the country: corn, beans and soybeans. While the impact is expected to vary across the country's different zones, the end results are expected to be greater evaporation and an increased need of water for irrigation of crops, a longer duration of the vegetative cycle, and reduced plant productivity. In the moderate temperature change scenario, the estimated fall for corn production is 5 percent to 30 percent, for beans 5 percent to 32 percent, and for soybeans 2.5 percent to 18 percent by the 2030.

Panama

Espinosa et al.[xlii] evaluated the impact of climate change on water resources in the La Villa, Chiriqui and Chagres river basins of Panama. The goal of the research was to develop different scenarios of water resource availability under given climate changes experienced by the doubling of global CO_2 concentration in the next 100 years.

For simulating impact, they used the model CLIRUN3 in combination with 20-year records of precipitation, potential evapotranspiration and water flow to simulate monthly river runoff in the Chagres (Panama Canal) river basin. This basin is critical because it supplies water to 25 percent of the country's population and is crucial to international navigation. The Chiriqui river basin is the main national source of hydropower and the La Villa river basin is highly important to agricultural activity. The Chagres river basin is part of the Atlantic watershed; the other two belong to the Pacific watershed.

PUERTO RICO—CLIMATE CHANGE IMPACTS ON WATER AVAILABILITY FOR A BIOENERGY PROJECT IN THE LAJAS VALLEY

Puerto Rico is looking at the feasibility of finding green energy alternatives. Researchers Guindin, Weiss and Pérez-Alegría[xliii] evaluated a bioenergy project based on sugarcane ethanol to use over 24,281 ha (60,000 ac) of prime farmland in the Lajas Valley.

Predicting sugarcane water needs under current conditions and for the future was considered a critical issue for the sustainability of any agricultural enterprise in the proposed region. In this research, the authors noted that there is intense competition for a finite amount of water among agricultural, residential, and commercial users. The objective of their effort was to study the impact of irrigation requirements for sugarcane using different climate change scenarios. The authors used the climate scenarios for 2010-2039 (2010s), 2040-2069 (2040s), and 2070-2099 (2070s) periods from the HadCM3 A21 model developed at the UK Hadley Climate Research Center and the CGM2 A21 model developed by the Canadian Climate Centre. Climate change scenarios were generated based on projections from these models. The relative changes in precipitation, maximum and minimum temperature were calculated for the three periods (2010s, 2040s, and 2070s) using the climate change scenarios from the HadCM3 and CGM2 models. Sugarcane water requirements were calculated with Crop Wat 4 using generated monthly temperature and precipitation for the three periods.

The authors state their conclusions as follows:

"Both climate change scenarios project a decrease in total annual precipitation for 2010s, 2040s and 2070s. The HadCM3 model projected a 43 mm decrease in total annual precipitation for 2010s while the CGM2 model projected a decrease of 400 mm for the same period. For 2070s, the HadCM3 model projected a 422 mm decrease in total annual precipitation. Under the current climate conditions, simulation results indicate that the irrigation system does not have the capacity to supply the irrigation water requirements for 60,000 acres of sugarcane in the Lajas Valley. Future irrigation water requirements for sugarcane show an increase over 90 percent under climate change scenarios for the periods 2010s, 2040s and 2070s, based on the actual irrigation system capacity. If the assumptions used in this study are reasonable, now is the time for planning future water supply and storage systems and developing alternatives crops that can adapt to less water. Further research is needed to assess other sources of uncertainty—in particular, changes in wet and dry periods, and to analyze the possible impact on other crops grown in the region."

The authors ran the model for the watersheds under scenarios with temperature increases of 1°C and 2°C, with precipitation changes of plus or minus 15 percent for the Pacific and plus or minus 20 percent for the Atlantic watershed. Although the model and information had limitations, the simulated results had a high correlation, 0.9, with the observed data. The simulation study showed "A clear indication that basins located in the Pacific region would be the most affected under the conditions of the incremental scenarios used." During November-December, when water demand is higher, water flow is projected to lessen as temperature increases, whether or not precipitation increases. This suggests that the basins are highly sensitive to temperature changes, particularly during the dry season.

Under a scenario of increased temperature and decreased precipitation, the mean monthly flow tends to decrease by 3 to 42 percent, both in the Atlantic and Pacific basins.

If simultaneous increases in temperature and precipitation took place, the flow in the Pacific basins would be reduced by 5 percent to 35 percent from November to March. During the remaining months the mean flow would increase by 4 percent to 40 percent. However, in the basin of the Atlantic watershed all the simulated values would be 3 percent to 50 percent higher than the mean value.

Espinosa et al. point out that there is great uncertainty in the assessment of changes in climatic conditions for different time periods because GCMs are not highly reliable tools for studies in the Central American region. However, the use of incremental scenarios allowed evaluation of how sensitive water resource availability is under different temperature increments and precipitation changes. (U

Adaptive Capacity

The impact of climate change on a society will be felt by how well it can adapt to climate change, that is, its adaptive capacity. Adaptive capacity is defined by the IPCC as, "The ability of a system to adjust to climate change (including climate variability and extremes), to moderate potential damages, to take advantage of opportunities, or to cope with the consequences."[xliv] Thus, adaptive capacity is distinguished from both the effects of climate change and the degree to which those effects influence the systems that are in place, as noted in the previous sections.

Although the specific determinants of adaptive capacity are a matter of debate among researchers, there is broad agreement that economic, human, and environmental resources are essential elements. Some components of this adaptive capacity are near-term, such as the ability to deliver aid swiftly to those affected by flooding or droughts for example. Other components include a high enough level of education so that people can change livelihoods, a quantity of unmanaged land that can be brought into food production, and institutions that provide knowledge and assistance in times of change. For instance, Yohe and Tol[xlv] have identified eight qualitative "determinants of adaptive capacity," many of which are societal in character, although the scientists draw on an economic vocabulary and framing:

1. The range of available technological options for adaptation.
2. The availability of resources and their distribution across the population.
3. The structure of critical institutions, the derivative allocation of decision-making authority, and the decision criteria that would be employed.
4. The stock of human capital, including education and personal security.
5. The stock of social capital, including the definition of property rights.
6. The system's access to risk-spreading processes.
7. The ability of decision-makers to manage information, the processes by which these decision-makers determine which information is credible, and the credibility of the decision- makers themselves.
8. The public's perceived attribution of the source of stress and the significance of exposure to its local manifestations.

The Caribbean and Central American Region in a Global Context

Researchers have only recently taken on the challenge of assessing adaptive capacity in a comparative, quantitative framework. A global comparative study of resilience to climate change, including adaptive capacity, was conducted using the Vulnerability-Resilience Indicators Model (VRIM—see description in box).[xlvi]

VULNERABILITY-RESILIENCE INDICATORS MODEL (VRIM)

The VRIM is a hierarchical model with four levels. The vulnerability index (level 1) is derived from two indicators (level 2): sensitivity (how systems could be damaged by climate change) and adaptive capacity (the capability of a society to maintain, minimize loss of, or maximize gains in welfare). Sensitivity and adaptive capacity, in turn, are composed of sectors (level 3). For adaptive capacity these sectors are human resources, economic capacity, and environmental capacity. For sensitivity, the sectors are settlement/infrastructure, food security, ecosystems, human health, and water resources. Each of these sectors is composed of one to three proxies (level 4). The proxies under adaptive capacity are as follows: human resource proxies are the dependency ratio and literacy rate; economic capacity proxies are GDP (market) per capita and income equity; and environmental capacity proxies are population density, sulfur dioxide divided by state area, and percent of unmanaged land. Proxies in the sensitivity sectors are water availability, fertilizer use per agricultural land area, percent of managed land, life expectancy, birthrate, protein demand, cereal production per agricultural land area, sanitation access, access to safe drinking water, and population at risk due to sea level rise.

Each of the hierarchical level values is comprised of the geometric means of participating values. Proxy values are indexed by determining their location within the range of proxy values over all countries or states. The final calculation of resilience is the geometric mean of all eight sectors.

Adaptive capacity as assessed in that study consists of seven variables, in three sectors, chosen to represent societal characteristics important to a country's ability to cope with and adapt to climate change:

Human and Civic Resources

- *Dependency Ratio:* proxy for social and economic resources available for adaptation after meeting basic needs.
- *Literacy:* proxy for human capital generally, especially the ability to adapt by changing employment.

Economic Capacity

- *GDP (market) Per Capita:* proxy for economic well-being in general, especially access to markets, technology, and other resources useful for adaptation.
- *Income Equity:* proxy for the potential of all people in a country or state to participate in the economic benefits available.
- Percent of Land that is Unmanaged: proxy for potential for economic use or increased crop productivity and for ecosystem health (e.g., ability of plants and animals to migrate under climate change).
- Sulfur Dioxide Per Unit Land Area: proxy for air quality and, through acid deposition, other stresses on ecosystems.
- Population Density: proxy for population pressures on ecosystems (e.g., adequate food production for a given population).

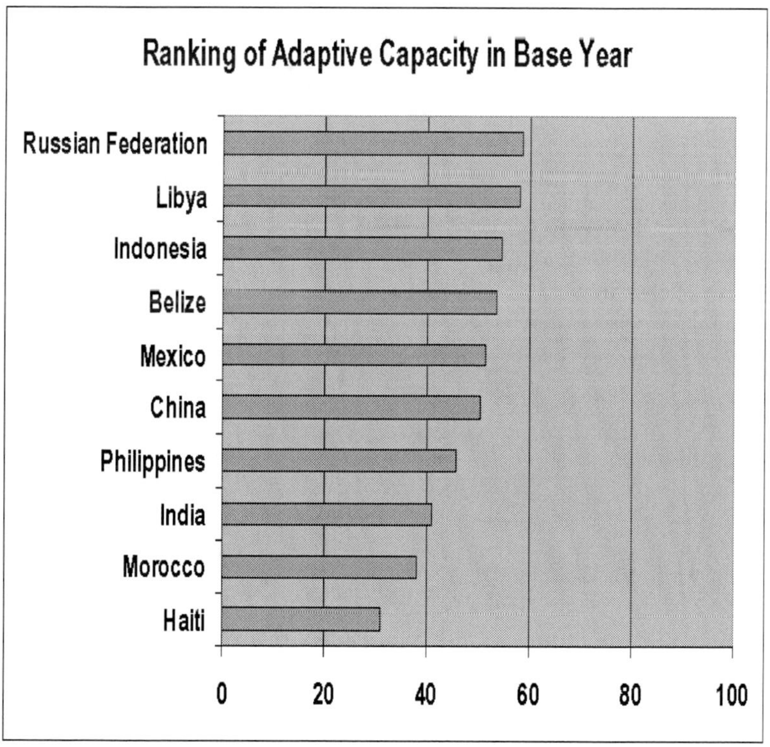

Figure 18. Sample of 10 countries' rankings of adaptive capacity (2000)

Adaptive capacity for a sample of 10 countries from the 160-country study is shown in Figure 18 (base year of 2000). There is a wide range of adaptive capacity represented by these countries; the three countries from the Caribbean—Belize, Mexico, and Haiti—are in the high-middle and lowest ranks, both in the sample and overall:

- Russia ranks 32nd and Libya 34th (in the highest quartile).
- Indonesia ranks 45th, Belize 48th, Mexico 59th , and China 75th (in the second quartile).
- The Philippines ranks 91st and India 119th (in the third quartile).
- Morocco ranks 136th and Haiti 156th (in the lowest quartile).

Any country-level analysis must take into account the comparative ranking of the country in the overall 160 groups of countries.

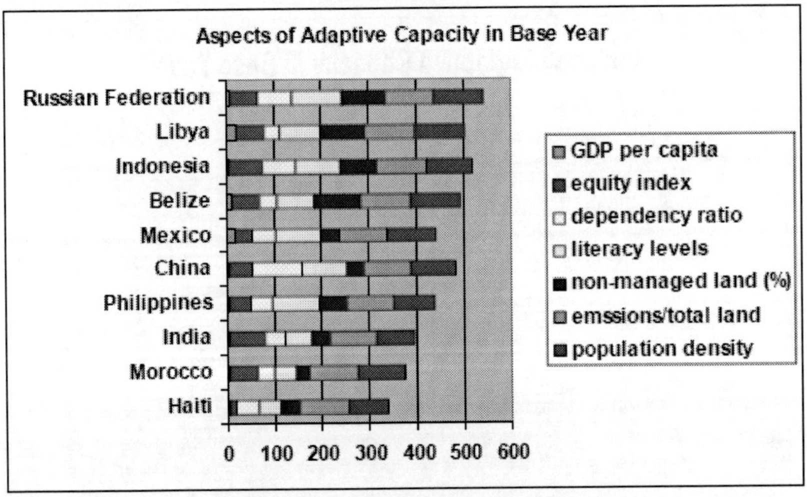

Figure 19. Variables' contributions to adaptive capacity rankings

Figure 19 shows the contribution of each variable to the overall ranking with slight differences occurring because of the methodology (see box above). Belize ranks fairly high because of favorable environmental capacity proxies (comparatively high percentage of unmanaged land, low emissions, and low population density). Mexico also ranks in the second quartile of countries overall, but with different strengths: in human and civic resources (comparatively favorable dependency and literacy levels) as well as environmental capacity (low emissions and low population density—but a less favorable percentage of unmanaged land). Haiti ranks poorly on almost every proxy variable, with the exception of emissions, which are comparatively low.

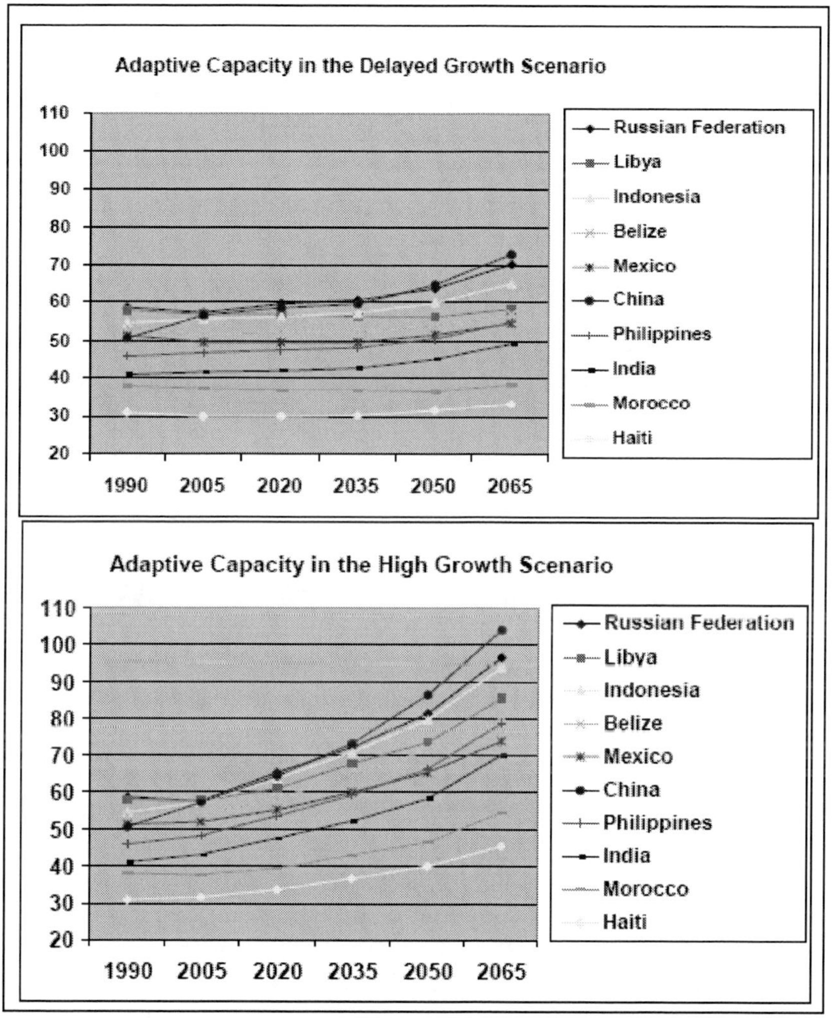

Source: Based on E.L. Malone and A.L. Brenkert, "Vulnerability, sensitivity, and coping/adaptive capacity worldwide," *The Distributional Effects of Climate Change: Social and Economic Implications*, M. Ruth and M. Ibarraran, eds., Elsevier Science, Dordrecht (in press).

Figure 20. Projections of adaptive capacity for 11 countries under a delayed growth scenario and a high growth scenario

Figure 20 shows projected adaptive capacity growth over time for the 10-country sample. Projections are made for two scenarios: rates of growth are based on the IPCC's A1 scenario in its Special Report on Emissions Scenarios, the A2A1 (delayed growth) and the A1v2 (high growth) scenario as adapted from the IPCC A1 and A2 scenarios by the IPCC participating model (MiniCAM). Both scenarios (A2A1 and A1v2) feature moderate population growth and a tendency toward convergence in affluence (with market-based solutions, rapid technological progress, and improving human welfare).

The scenarios used in this study differ in the rate of economic growth, one modeling high-andfast economic growth and the other delayed growth. In the delayed-growth scenario, the three Caribbean and Central American countries show almost stagnant, then modest growth. In the high-growth scenario, all countries improve their adaptive capacity, although the overall gap among different countries widens (i.e., initially lower-ranking countries do not show as high growth rates as initially higher-ranking countries). Both scenarios show the Philippines improving its adaptive capacity at a higher rate than Mexico and, in the high-growth scenario, overtaking Mexico.

Caribbean and Central American Countries Compared to Each Other

Turning to the specific set of countries included in this report,[8] Figure 21 shows the base year values by sector and by proxy variable for all nine Central American and Caribbean countries. Here the differences among countries in elements of adaptive capacities are clear, e.g., human resources strengths in Panama and Cuba, environmental capacity strengths in Belize and Honduras.

For projections, Figure 22 provides two snapshots of the low-growth scenario (A2A1) for 2020 and 2050. In both future years, many countries are projected to experience a decrease in environmental capacity, in some cases partially compensated for by increases in other areas (e.g., GDP per capita in Belize and Mexico; literacy levels in Panama, Cuba, and the Dominican Republic).

---

[8] Except Puerto Rico.

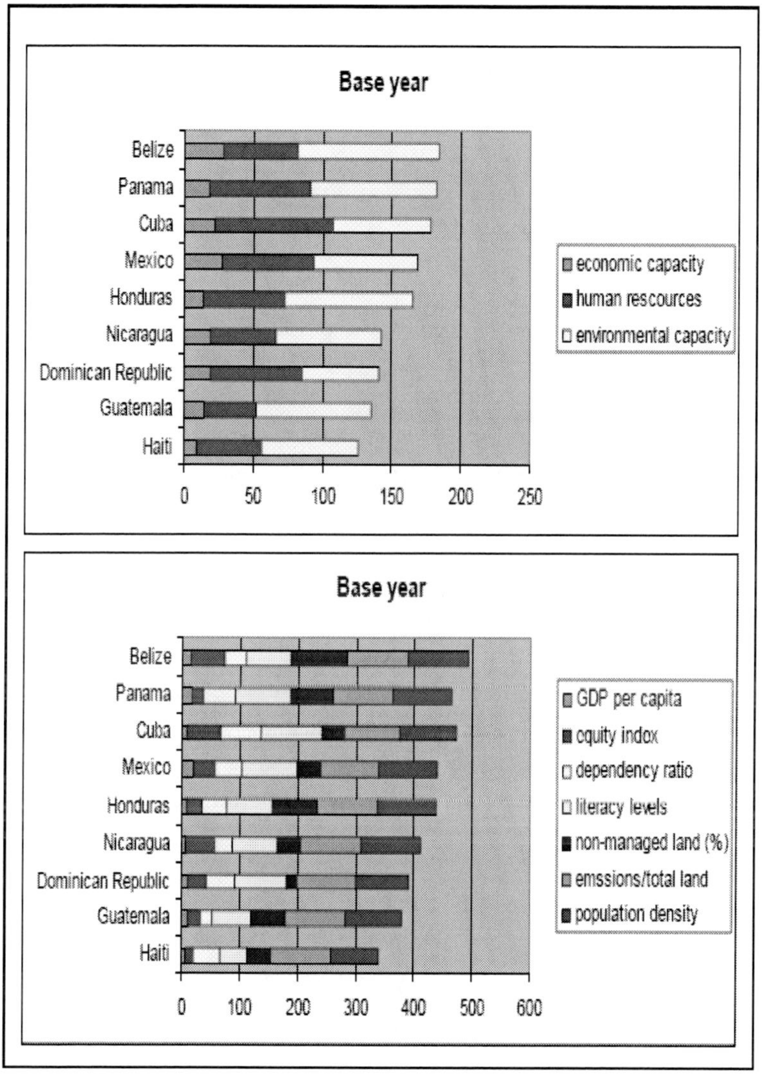

Source: Based on E.L. Malone and A.L. Brenkert, "Vulnerability, sensitivity, and coping/adaptive capacity worldwide," *The Distributional Effects of Climate Change: Social and Economic Implications, M. Ruth and M. Ibarraran*, eds., Elsevier Science, Dordrecht (in press).

Figure 21. Base year rankings of adaptive capacity in nine Caribbean/Central American countries

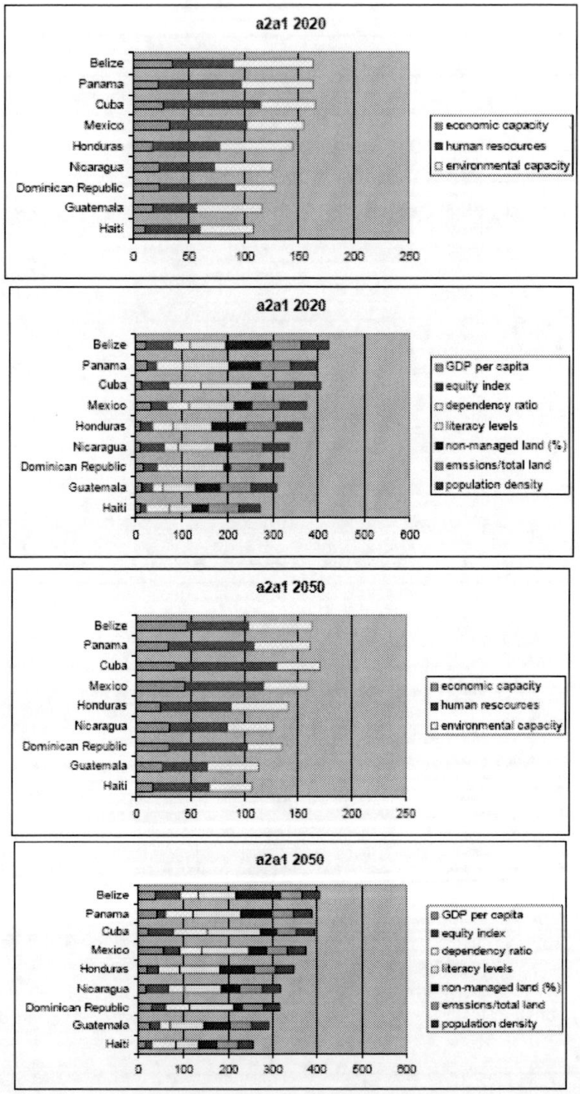

Source: Based on E.L. Malone and A.L. Brenkert, "Vulnerability, sensitivity, and coping/adaptive capacity worldwide," *The Distributional Effects of Climate Change: Social and Economic Implications*, M. Ruth and M. Ibarraran, eds., Elsevier Science, Dordrecht (in press)

Figure 22. 2020 and 2050 snapshots of the low-growth scenario for Caribbean/Central American countries, with sector results and proxy variable results

Key Contributors to Adaptive Capacity by Country

As stated above, there are several key indicators/parameters for any given country that can provide insight into its adaptive capacity, such as literacy rates, basic services, energy supply, and changes in production. In Latin America and the Caribbean, population has steadily increased since the 1900s and is expected to continue the trend through 2030. Availability of adequate human resources is a necessary condition to enhance adaptive capacity. It is also important that these resources have the appropriate level of education and access to basic services in order to have the ability to support economic growth.

Illiteracy continues to be a concern in some of the countries of interest. An illiterate person is defined as an individual unable to read and write a short simple statement on his or her everyday life. Significant progress has been made in most countries of the region. Nicaragua and Haiti, however, in 2005 still had greater than one third of the population older than 15 years of age classified as illiterate. This significantly affects economic growth, economic diversification, and adaptive capacity. Table 8 shows past and projected population for the selected countries, and Table 9 illustrates the level of illiteracy in the region.

Throughout the 1970s, 1980s, and 1990s the countries in Central America and the Caribbean experienced long periods of social unrest, capital flight, economic contraction, and large intraregional and extra-regional migration. In many cases the best educated members of a population emigrated. Intra-regional migration during these three decades grew rapidly. Nicaragua and El Salvador, in particular, saw many of their best flee to Costa Rica beginning in the 1 970s, and by 2000 over 8 percent of Costa Rica consisted of immigrants from those two countries. This was the direct result of the civil wars fought in both countries.

During the same period, Mexico also received many migrants from Guatemala and Nicaragua. At the end of the 1990s, Guatemala and the other countries in the region signed peace agreements and experienced the repatriation of many of their citizens from Mexico. By 2000, Mexico had a significantly smaller portion of immigrants from these countries than it had in 1990. There is also the added element of intra-regional seasonal migration exercised by those following the agricultural sector for employment. On a yearly basis, there are migrations from northern Panama to southern Costa Rica and from northern Guatemala to southern Mexico.

Migration from Central America and the Caribbean to the United States also increased during the same period. Caribbean-born immigrants accounted for almost 10 percent of the total US foreign-born population in 2000. The largest growth in the number of immigrants from Latin America to the United States occurred from 1990 to 2000 when a 97 percent increase occurred. Population grew from 7.2 million to 14.2 million. The 14.2 million people in 2000 included 9.1 million from Mexico, 879,000 from Cuba, 710,000 from Dominican Republic, 468,000 from Guatemala, 409,000 from Haiti, and 232,000 from Nicaragua. The proportion of the original total population that migrated to the United States during this period represented a wide range of the total population in the country of origin in the year 2000. The proportion ranged from 13 percent in the case of El Salvador, 9 percent for Mexico, 8 percent for Dominican Republic, 7.8 percent for Cuba, and 4 percent each for Haiti and Nicaragua.

Table 8. Total Population (Thousands)

Country	1990	2000	2010	2030
Belice	186.0	245.0	306.0	413.0
Costa Rica	3076.0	3925.0	4695.0	5779.0
Cuba	10605.0	11129.0	11236.0	11 077.0
El Salvador	5 110.0	6 276.0	7 453.0	9 652.0
Guatemala	8908.0	11225.0	14362.0	21 804.0
Haití	7108.0	8576.0	10085.0	13 350.0
Honduras	4901.0	6231.0	7614.0	10 414.0
México	84002.0	99684.0	110056.0	127 211.0
Nicaragua	4141.0	5106.0	5825.0	7 140.0
Puerto Rico	3528.0	3834.0	4056.0	4 383.0
República Dominicana	7296.0	8740.0	10169.0	12 625.0

Source: CEPAL/ECLAC [Comisión Económica para América Latina y el Caribe/Economic Commission for Latin America and the Caribbean], *Anuario Estadístico de América Latina y el Caribe: 2008* (United Nations 2009)

The migration from Central America and the Caribbean, intra-regional and extra regional, has resulted in a systematic and regular transfer of funds from the United States and other countries to the families and relatives that remained in the countries of origin. The Inter-American Development Bank (IADB) estimates the region received US$7.8 billion through official channels in 2004, a 17 percent increase from the 2003 figure of US$6.7 billion.[xlvii]

Table 9. Percentage of Illiterate Population (15 years or older)

	Both sexes				
Country	1970	1980	1990	2000	2005
Belice	25.0	17.5	10.9	6.8	5.3
Costa Rica	11.8	8.3	6.1	4.4	3.8
Cuba	10.7	7.5	4.9	3.3	2.7
El Salvador	42.1	34.2	27.6	21.3	18.9
Haití	78.0	69.5	60.3	50.2	45.2
Honduras	49.4	40.1	31.9	25.0	22.0
México	26.5	18.7	12.7	8.8	7.4
Nicaragua	45.5	41.2	37.3	33.5	31.9
Puerto Rico	14.7	11.1	8.5	6.2	5.4
República Dominicana	32.8	26.0	20.6	16.3	14.5

Source: CEPAL/ECLAC [Comisión Económica para América Latina y el Caribe/Economic Commission for Latin America and the Caribbean], Anuario Estadístico de América Latina y el Caribe: 2008 (United Nations 2009).

Guatemala topped the list of recipients with almost US$2.7 billion in official flows in 2004, followed by El Salvador with US$2.5 billion. These two countries, which account for nearly two-thirds of the two million Central Americans counted in the 2000 US census, receive almost 64 percent of total remittance flows to Central America. They are the fourth- and fifth-largest remittance-receiving countries in Latin America and the Caribbean. Remittance growth in Guatemala tripled from 2001 to 2004. Honduras and Nicaragua followed at some distance, at around the US$1 billion, while Panama, Costa Rica, and Belize trailed with less than US$325 million in remittances in 2004. The low levels of the latter three reflect the fact that they have relatively few emigrants in the United States.

While much attention is given to remittances from developed countries, particularly the United States, there are substantial intra-regional remittance flows too. A 2003 study of Costa Rica and Nicaragua revealed that about one-third of remittances received in Nicaragua are sent from Costa Rica. Since Mexico is the second largest destination of Guatemalan workers after the United States, it can easily be concluded that some of the remittances going to Guatemala are coming from Mexico. Research conducted for IADB estimated that in 2002 about US $1.5 billion of the US$32 billion remitted to Latin America and the Caribbean were actually intra-regional.

Another key indicator of the level of adaptive capacity is the infrastructure for basic services. In most of the countries selected for this assessment the majority of the population is concentrated in urban areas, and the largest urban areas are found in the coastal areas of the countries. Basic infrastructure/ services such as water, electricity, and sewage are important elements in the ability to reduce and recover from the impact of such extreme events as hurricanes, floods, and droughts. Table 3 depicts the level of basic infrastructure in some of the selected countries.

Another key contributor to adaptive capacity is the extent of forests in this region. Deforestation is a significant environmental issue for every country selected for this report. Puerto Rico does not suffer from this environmental problem. According to the IPCC assessment,[xlviii] by 2010 the forest areas in Central America will be reduced by 1.2 Mha. These areas are projected to be used for pasture and expanding livestock production. Table 10 illustrates the loss in forest area by country from 1990 to 2005.

During this timeframe Cuba was the only country that experienced increases in forest area. Except for the Dominican Republic, which maintained the size of its forest area, all the other countries have steadily reduced their forests, from 6 percent in Costa Rica and Mexico to 37 percent in Honduras.

Several countries in Central America and the Caribbean as well as Mexico have made an effort to increase the amount of protected areas. Table 11 shows how the selected countries have changed protected areas from 1990 to 2007. Mexico is the largest contributor, having doubled the amount of land under protection and increasing the amount of marine areas many-fold during the same period.

Table 10. Forest area and proportion of land area covered by forest

(Thousands of hectares, percentage and rate of variation)												
	Forest area (Thousands of hectares)			covered by forest (Percentage)			Accumulated variation in forest area (Rate of variation)			Average annual variation in forest area (Rate of variation)		
	1990	2000	2005	1990	2000	2005	1990-2000	2000-2005	1990-2005	1990-2000	2000-2005	1990-2005
Belice	1 653	1 653	1 653	72.5	72.5	72.5	-	-	-	-	-	-
Costa Rica	2 564	2 376	2 391	50.2	46.5	46.8	-7.3	0.6	-6.7	-0.7	0.1	-0.4
Cuba	2 058	2 435	2 713	18.7	22.2	24.7	18.3	11.4	31.8	1.8	2.3	2.1
El Salvador	375	324	298	18.1	15.6	14.4	-13.6	-8.0	-20.5	-1.4	-1.6	-1.4
Guatemala	4 748	4 208	3 938	43.8	38.8	36.3	-11.4	-6.4	-17.1	-1.1	-1.3	-1.1
Haiti	116	109	105	4.2	4.0	3.8	-6.0	-3.7	-9.5	-0.6	-0.7	-0.6
Honduras	7 385	5 430	4 648	66.0	48.5	41.5	-26.5	-14.4	-37.1	-2.6	-2.9	-2.5
México	69 016	65 540	64 238	35.5	33.7	33.0	-5.0	-2.0	-6.9	-0.5	-0.4	-0.5
Nicaragua	6 538	5 539	5 189	53.9	45.6	42.7	-15.3	-6.3	-20.6	-1.5	-1.3	-1.4
República Dominicana	1 376	1 376	1 376	28.4	28.4	28.4	*	*	*	*	*	*
América Latina y el Caribe	**984 123**	**939 208**	**915 494**	**49.1**	**46.8**	**45.6**	**-4.6**	**-2.5**	**-7.0**	**-0.5**	**-0.5**	**-0.5**

Note: no data for Panama or Puerto Rico. Source: CEPAL/ECLAC [Comisión Económica para América Latina y el Caribe/Economic Commission for Latin America and the Caribbean], *Anuario Estadístico de América Latina y el Caribe: 2008* (United Nations 2009)

Table 11. Protected Areas in Selected Latin American and Caribbean Countries

	TERRESTRIAL PROTECTED AREAS (Square kilometers)					*MARINE PROTE CTED AREAS (Square kilometers)*				
País	**1990**	**2000**	**2005**	**2006**	**2007**	**1990**	**2000**	**2005**	**2006**	**2007**
Belice	...	8912	7940	7944	8006	...	1587	2498	2498	2498
Costa Rica	...	12 755	13 333	13 266	13 558	...	...	...	5 209	5 210
Cuba	...	...	3304	3 309	3309	...	...	2 049	2071	2071
El Salvador	...	...	...	...	416	...	...	...	...	206
Guatemala	25107	31 180	33077	30 801	30 890	158	158	158	2453	2453
Honduras	...	...	28 821	31636	...	...	...	665	1155	...
México	76640	131775	148505	180210	187004	4 408	35255	40660	40660	45021
República Dominicana	...	9 176	10529	10529	10529	...	17494	67602	67602	...

Note: no data for Haiti, Nicaragua, Panama, and Puerto Rico

Source: CEPAL/ECLAC [Comisión Económica para América Latina y el Caribe/Economic Commission for Latin America and the Caribbean], *Anuario Estadístico de América Latina y el Caribe: 2008* (United Nations 2009).

UNDP and ECLAC are beginning a series of studies to quantify the impact of climate change in socio-economic and ecosystems in the region. Even if these studies are not yet available, leaders in the region now accept that, while the region does not contribute to global greenhouse gases in a significant way, it is highly vulnerable to the effects generated by severe climate variability.

This has been observed over the past 20 years, and leaders understand that it is critical for them to develop sustainable development policies and to enhance their capabilities to respond and adapt to severe weather events.

Energy. Energy resources, production, and use vary widely across the countries under review. All the countries under review will experience population growth, economic growth, and industrialization, they will increase their need and demand for energy. All the countries rely on imported fossil fuels, with the exception of Mexico, which is a net exporter of energy resources. In most countries the largest generator of greenhouse gases is the energy sector. Although they are very small contributors to global emissions, most countries will benefit from increasing use of renewable energy. Most have begun efforts to evaluate and implement small projects, such as wind energy in Nicaragua and Costa Rica and an intensive effort in the Dominican Republic to evaluate hydro electricity.

Agriculture. The agricultural sector climate related research for most of the countries in this review is limited. Where research is available, productivity losses are projected for optimist, moderate, and pessimist scenarios for some key food crops with estimates that vary from 10 percent to more than 50 percent by the year 2030.

Water Resources. The majority of the population in most of the countries reviewed lives in coastal areas, which are highly vulnerable to severe climate changes. As populations continue to grow in the same areas, increasing water extraction and rising sea levels are expected to have severe impact on the quantity and quality of water available. Many of the aquifers of these countries are open to ocean waters and are already experiencing increases in salinity. Rising sea levels will accelerate the deterioration of aquifers and available water resources.

Migration. An increase in intra-regional and extra-regional migration during the 1 980s and 1 990s resulted from social unrest and economic contraction. Moreover, the inability of countries in the region to adapt and recover from severe climate events with major impacts on their economies will continue to promote migration outside the region, in particular, to the United States and Canada. The large number of immigrants coming to the United States in the past 20- 25 years will facilitate this movement.

In addition, the observed and projected incidence of diseases and pathogens varies across the countries under review. In Central American countries, there has been a sharp increase in the number of diseases during the years following ENSO effects. The Government of the Dominican Republic has not observed and has not projected a correlation between climate change variability and increases in health effects of its population. It is not clear if it is a difference in the quality of information or the limitations of the models used in the initial assessments of each country.

Although most countries in the Central America and Caribbean region have started to evaluate the impact of climate change in their economic, social, and natural resources, there is limited understanding of the viable options to address the problems.

Many limitations that exist today on climate change preclude making projections good enough to take action. They include limitations in models used, quality of data, and quantity of relevant data. Equally problematic is the limitation of funding to undertake detailed modeling for each country in such a way that the result is information that also ranks, evaluates and recommends financial options.

Although the countries under review have submitted their First National Communications to the UNFCC (and Mexico has submitted its third communication), significant work and analysis remain. Reviewers must still capture the full impact on socio-economic systems and the ability of those systems to recover and adapt to and reduce the effects of severe weather events.

The first assessments submitted by these countries have laid the foundation for improving models used and for improving the quality and quantity of data. The initial studies have also illustrated the gaps that exist between the current level of knowledge and what is needed for the development of policies that will improve the adaptive and response capacities of the countries under review.

ANNEX A: ACCURACY OF REGIONAL MODELS

This is an excerpt from IPCC (2007), Chapter 11, Regional models; see IPCC 2007 for references.[9]

11.6.2. Skill of Models in Simulating Present Climate

In the Central America (CAM) and Amazonia (AMZ) regions, most models in the multi-model dataset (MMD) have a cold bias of 0°C to 3°C, except in AMZ in September, October, and November (SON). In southern South America (SSA) average biases are close to zero. The biases are unevenly geographically distributed. The MMD mean climate shows a warm bias around 30°S (particularly in summer) and in parts of central South America (especially in SON). Over the rest of South America (central and northern Andes, eastern Brazil, Patagonia) the biases tend to be predominantly negative. The SST biases along the western coasts of South America are likely related to weakness in oceanic upwelling.

For the CAM region, the multi-model scatter in precipitation is substantial, but half of the models lie in the range of –15 to 25 percent in the annual mean. The largest biases occur during the boreal winter and spring seasons, when precipitation is meager. For both AMZ and SSA, the ensemble annual mean climate exhibits drier than observed conditions, with about 60 percent of the models having a negative bias. Unfortunately, this choice of regions for averaging is particularly misleading for South America since it does not clearly bring out critical regional biases such as those related to rainfall underestimation in the Amazon and La Plata Basins. Simulation of the regional climate is seriously affected by model deficiencies at low latitudes. In particular, the MMD ensemble tends to depict a relatively weak Inter-Tropical Convergence Zone (ITCZ), which extends southward of its observed position. The simulations have a systematic bias towards underestimated rainfall over the Amazon Basin. The simulated subtropical climate is typically also adversely affected by a dry bias over most of south-eastern South America and in the South Atlantic Convergence Zone, especially during the rainy season. In contrast, rainfall along the Andes and in northeast Brazil is excessive in the ensemble mean.

---

[9] Some references in this section have been changed to be internally consistent with this document and other references have been removed to avoid confusion.

Some aspects of the simulation of tropical climate with AOGCMs have improved. However, in general, the largest errors are found where the annual cycle is weakest, such as over tropical South America (see, e.g., Section 8.3). Atmospheric GCMs approximate the spatial distribution of precipitation over the tropical Americas, but they do not correctly reproduce the temporal evolution of the annual cycle in precipitation, specifically the mid-summer drought (Magaña and Caetano, 2005). Tropical cyclones are important contributors to precipitation in the region. If close to the continent, they will produce large amounts of precipitation over land, and if far from the coast, moisture divergence over the continental region enhances drier conditions.

Zhou and Lau (2002) analyse the precipitation and circulation biases in a set of six AGCMs provided by the Climate Variability and Predictability Programme (CLIVAR) Asian-Australian Monsoon AGCM Intercomparison Project (Kang et al., 2002). This model ensemble captures some large-scale features of the South American monsoon system reasonably well, including the seasonal migration of monsoon rainfall and the rainfall associated with the South America Convergence Zone. However, the South Atlantic subtropical high and the Amazonia low are too strong, whereas low-level flow tends to be too strong during austral summer and too weak during austral winter. The model ensemble captures the Pacific-South American pattern quite well, but its amplitude is generally underestimated.

Regional climate models (RCMs) are still being tested and developed for this region. Relatively few studies using RCMs for Central and South America exist, and those that do are constrained by short simulation length. Some studies (Chou et al., 2000; Nobre et al., 2001; Druyan et al., 2002) examine the skill of experimental dynamic downscaling of seasonal predictions over Brazil. Results suggest that both more realistic GCM forcing and improvements in the RCMs are needed. Seth and Rojas (2003) performed seasonal integrations driven by reanalyses, with emphasis on tropical South America. The model was able to simulate the different rainfall anomalies and large-scale circulations but, as a result of weak low-level moisture transport from the Atlantic, rainfall over the western Amazon was underestimated. Vernekar et al. (2003) follow a similar approach to study the low-level jets and report that the RCM produces better regional circulation details than does the reanalysis. However, an ensemble of four RCMs did not provide a noticeable improvement in precipitation over the driving large-scale reanalyses (Roads et al., 2003).

Other studies (Misra et al., 2003; Rojas and Seth, 2003) analyze seasonal RCM simulations driven by AGCM simulations. Relative to the AGCMs, regional models generally improve the rainfall simulation and the tropospheric circulation over both tropical and subtropical South America. However, AGCM-driven RCMs degrade compared with the reanalyses-driven integrations and they could even exacerbate the dry bias over sectors of AMZ and perpetuate the erroneous ITCZ over the neighboring ocean basins from the AGCMs. Menéndez et al. (2001) used a RCM driven by a stretched-grid AGCM with higher resolution over the southern mid- latitudes to simulate the winter climatology of SSA. They find that both the AGCM and the regional model have similar systematic errors but the biases are reduced in the RCM. Analogously, other RCM simulations for SSA give too little precipitation over the subtropical plains and too much over elevated terrain (e.g., Nicolini et al., 2002; Menéndez et al., 2004).

Annex B. Information Deficiencies that Preclude a Full Evaluation of the Impact of Climate Change on Central America, the Caribbean, and the Region's Adaptive Capacity

Regional leaders have not addressed the problem of the projected impact of climate change with possible policy changes or infrastructure investments because of a lack of systematic analysis that quantifies and qualifies the potential impact to the region. This lack of rigorous analysis restricts the development of relevant and economically viable options. There are significant gaps in the ability to fully understand all the dimensions of climate change at the economic, social, and/or environmental level in the region in a systematic way. There are gaps and deficiencies in data, systematic methodologies/analysis, and tools to monitor, share, and track information and events at the local, national, and regional levels. Efforts are starting to be made to reduce these gaps. Several entities at the national and regional levels are working to develop better analytical methods and information-sharing as well as better data and availability.

To increase the likelihood that this evaluation represents a reasonable assessment of projected climate change and its impact in Central America and the Caribbean as well as the region's adaptive capacity, the following gaps would need to be addressed:

- In physical science research, regional analyses will continue to be limited by the inability to model regional climates satisfactorily, including complexities arising from the interaction of global, regional, and local processes. Uncertainties in the occurrence and impact of the ENSO phenomenon, hurricane activity, and storm surges for example leave important gaps in knowledge needed for climate projections. One gap of particular interest is the lack of medium-term (20-30 years) projections that could be relied upon for planning purposes. Similarly, scientific projections of water supply and agricultural productivity are limited by inadequate understanding of various climate and physical factors affecting both areas. Research agendas in these areas can be found in the synthesis and assessment reports of the US Climate Change Science Program (*http://www.climate*science.gov) for instance and the National Academy of Sciences (e.g., *http://books.nap.edu/catalog.php? record_id=11175#toc)*. Similar types of issues exist for the biological and ecological systems that are affected.

- In social science research, scientists and analysts have only partial understandings of the important factors in vulnerability, resilience, and adaptive capacity, much less their interactions and evolution. Again, research agendas on vulnerability, adaptation, and decision-making abound (e.g., (http://books.nap. edu/catalog.php?record_id=12545).

- Important factors are unaccounted for in research; scientists know what some of them are, but there are likely factors whose influence will be surprising. An example from earlier research on the carbon cycle illustrates this situation. The first carbon cycle models did not include carbon exchanges involving the terrestrial domain. Modelers assumed that the exchange was about equal, and the only factor modeled was deforestation. This assumption, of course, made the models inadequate for their purposes. In another example, ecosystems research models are only beginning to account for changes in pests, e.g., the pine bark beetle.

- Social models or parts of models in climate research have been developed to simulate consumption (with the assumption of well functioning markets and rational actor behavior) and mitigation/ adaptation policies (but without attention to the social feasibility of enacting or implementing such policies). As anthropogenic climate change is the result of human decisions, the lack of knowledge about motivation, intent, and behavior is a serious shortcoming.

Overall, research about the impact of climate change on the Central America and Caribbean region has been undertaken piecemeal: discipline by discipline, sector by sector, with political implications separately considered from physical effects. Outside the National Communications, small-scale case studies have been done, but little systematic analysis. This lack of rigorous analysis can be remedied by integrated research into the energy, economic, environmental, and political conditions and possibilities.

CONCLUSIONS

The systematic evaluation of the impact of climate change in the Caribbean and Central American is only beginning. There are many limitations associated with data quality and quantity. Most of the countries, however, are beginning to quantify greenhouse gas inventories and run simulation models to estimate the potential impact associated with projected global average increase in temperatures, rise in sea level, and changes in rainfall.

Annex C: Environmental Agreements Signed by Selected Latin American and Caribbean Countries

Multilateral Environmental Agreements

Year of signature and year that the country became party to the agreement (through ratification, acceptance, approval or adhesion)

Country	Ramsar /a/	Patrimonio /b/ Heritage	CITES /c/	Especies migratorias /d/ CMS	Derecho del mar Law of the Sea		Viena /e/ Vienna		Montreal /f/		Basilea /g/ Basel		Biological Diversity /h/		Cambio Climatico UNFCCC /i/	
	P	P	P	EV	F	P	F	P	F	P	F	P	F	P	F	P
Belice	1998	1990	1986	...	1982	1983	...	1997	...	1998	..	1997	1992	1993	1992	1994
Costa Rica	1991	1977	1975	2007	1982	1992	...	1991	...	1991	..	1995	1992	1994	1992	1994
Cuba	2001	1981	1990	2008	1982	1984	...	1992	...	1992	..	1994	1992	1994	1992	1994
El Salvador	1999	1991	1987	...	1984	...	...	1992	...	1992	1990	1991	1992	1994	1992	1995
Guatemala	1990	1979	1979	...	1983	1997	...	1987	...	1989	1989	1995	1992	1995	1992	1995
Haití	..	1980	..	...	1982	1996	...	2000	...	2000	1989	...	1992	1996	1992	1996
Honduras	1993	1979	1985	2007	1982	1993	...	1993	...	1993	..	1995	1992	1995	1992	1995
México	1986	1984	1991	...	1982	1983	1985	1987	1987	1988	1989	1991	1992	1993	1992	1993
Nicaragua	1997	1979	1977	...	1984	2000	...	1993	...	1993	..	1997	1992	1995	1992	1995
República Dominicana	2002	1985	1986	...	1982	...	...	1993	...	1993	..	1999	1992	1996	1992	1998

Country	Desertificación UNCCD /k		Kyoto /l		Rotterdam /m		Cartagena /n		Estocolmo Stockholm /o	
	F	P	F	P	F	P	F	P	F	P
Belice	...	1998	...	2003	...	2005	...	2004	2002	..
Costa Rica	1994	1998	1998	2002	1999	..	2000	2007	2002	2007
Cuba	1994	1997	1999	2002	1998	2008	2000	2002	2001	2007
El Salvador	...	1997	1998	1998	1999	1999	2000	2003	2001	2008
Guatemala	...	1998	1998	1999	...	..	...	2004	2002	2008
Haití	1994	1996	...	2005	...	..	2000	..	2001	..
Honduras	1995	1997	1999	2000	...	..	2000	..	2002	2005
México	1994	1995	1998	2000	...	2005	2000	2002	2001	2003
Nicaragua	1994	1998	1998	1999	...	2008	2000	2002	2001	2005
República Dominicana	...	1997	...	2002	...	2006	...	2006	2001	2007

Source: CEPAL- Anuario Estadistico, 2008

j/ UNFCCC: The United Nations Framework Convention on Climate Change, 1992.

k/ UNCCD: The United Nations Convention to Combat Desertification in Those Countries Experiencing Ser ious Drought and/or Desertification, 1994.

l/ Kyoto: The Kyoto Protocol to the United Nations Framework Convention on Climate Change, 1997.

m/ Rotterdam: The Rotterdam Convention on the Prior Informed Cons ent Procedure for Cartain Hazardous Chemicals and Pesticides in International Trade, 1998.

n/ Cartagena: The Car tagena Protocol on Biosafety to the Convention on Biological Diversity, 2000.

o/ Stockholm: The Stockholm Convention on Persistnt Organic Pollutants, 2001.

p/ The year that the countries signed the agreement is not available.

q/ All the countries that are party to this convention signed it between 1973 and 1974, the period in which the c onvention was open for signature.

End Notes

[i] Information for the following brief descriptions is from Central Intelligence Agency (CIA), The 2008 World Factbook, *https://www.cia.gov/library/publications/the-world-factbook/index.html* (accessed May 14, 2009).

[ii] Loretta Burke and Jonathan Maidens, "Haiti," in *Reefs at Risk in the Caribbean, World Resources Institute* (2004) http://www.wri.org/publication/content/7889.

[iii] J.E. Oliver, ed., "Central America and the Caribbean," in *Encyclopedia of World Climatology:* Encyclopedia of Earth Sciences Series (Dordrecht: Springer 2005).

[iv] J.J. McCarthy, O.F. Canziani, N.A. Leary, D.J. Dokken and K.S. White, eds., *Climate Change 2001: Impacts, Adaptation, and Vulnerability: Contribution of Working Group II to the Third Assessment Report of the Intergovernmental Panel on Climate Change* (Cambridge: Cambridge University Press 2001).

[v] T.C. Peterson et al., "Recent changes in climate extremes in the Caribbean region," *Journal of Geophysical Research 107* (2002): 4601, doi:10.1029/2002JD002251, 2002.

[vi] J.T. Houghton, D.A. Callander and S.K. Varney, eds., *Climate Change 1992: The Supplementary Report to the IPCC Scientific Assessment* (Cambridge: Cambridge University Press 1998).

[vii] M.L. Parry, O.F. Canziani, J.P. Palutikof, P.J. van der Linden and C.E. Hanson, eds., *Climate Change 2007: Impacts, Adaptation and Vulnerability: Contribution of Working Group II to the Fourth Assessment Report of the Intergovernmental Panel on Climate Change* (Cambridge: Cambridge University Press 2007).

[viii] J.D. Neelin, M. Münnich, H. Su, J.E. Meyerson and C.E. Holloway, "Tropical drying trends in global warming models and observations," *Proceedings of the National Academy of Sciences 103* (April 18, 2006): 16.

[ix] The Bahamas, "First National Communication on Climate Change to the Secretariat of the United Nations Framework Convention on Climate Change" (Nassau: The Bahamas Environment, Science and Technology Commission April 2001).

[x] The Bahamas, "First National Communication on Climate Change to the Secretariat of the United Nations Framework Convention on Climate Change" (Nassau: The Bahamas Environment, Science and Technology Commission April 2001).

[xi] J.A. Pounds, M.P.L. Fogden and J.H. Campbell, "Biological response to climate change on a tropical mountain," *Nature* 398(1999): 611 –615.

[xii] Intergovernmental Panel on Climate Change (IPCC), *Special Report on Emissions Scenarios (SRES),* eds. Nebojsa Nakicenovic and Rob Swart (Cambridge: Cambridge University Press, 2000), http://www.ipcc.ch/ipccreports/sres/emission

[xiii] J.H. Christensen and B. Hewitson, "Regional Climate Projections," in *Climate Change 2007: the Physical Science Basis*, eds. S. Solomon, D. Qin, M. Manning, M. Marquis, K. Averyt, M.M.B. Tignor, H.L. Miller Jr. and Z. Chen (Cambridge: Cambridge University Press 2007).

[xiv] J.H. Christensen and B. Hewitson, "Regional Climate Projections: Supplementary Material," in *Climate Change 2007: the Physical Science Basis*, eds. S. Solomon, D. Qin, M. Manning, M. Marquis, K. Averyt, M.M.B. Tignor, H.L. Miller Jr. and Z. Chen (Cambridge: Cambridge University Press 2007).

[xv] J.H. Christensen and B. Hewitson, "Regional Climate Projections," in *Climate Change 2007: the Physical Science Basis*, eds. S. Solomon, D. Qin, M. Manning, M. Marquis, K. Averyt, M.M.B. Tignor, H.L. Miller Jr. and Z. Chen (Cambridge: Cambridge University Press 2007).

[xvi] M. Angeles et al., "Predictions of future climate change in the Caribbean region using global general circulation models," *International Journal of Climatology* 27:5 (2007): 555-569.

[xvii] A.V. Karmalkar, R.S. Bradley and H.F. Diaz, "Projections of Future Changes in Precipitation for Central America Using PRECIS Regional Climate Model," (American Geophysical Union, Fall Meeting, San Francisco 2008).

[xviii] A.V. Karmalkar, R.S. Bradley and H.F. Diaz, "Climate change scenario for Costa Rican montane forests," *Geophysical Research Letters 35* (2008).

[xix] N. Nakicenovic and R. Swart, *Special Report on Emissions Scenarios* (Cambridge: Cambridge University Press 2000).

[xx] J.P. Kossin et al., *A globally consistent reanalysis of hurricane variability and trends, Geophysical Research Letters,* Vol. 34, L04815, doi:10.1029/2006GL028836, 2007

[xxi] K.E. Trenberth and D.J. Shea, "Atlantic hurricanes and natural variability in 2005," *Geophysical Research Letters 33* (2006): L12704, doi:10.1029/2006GL026894.

[xxii] K.E. Trenberth and D.J. Shea, "Atlantic hurricanes and natural variability in 2005," *Geophysical Research Letters 33* (2006): L12704, doi:10.1029/2006GL026894.

[xxiii] J. Leggett, W. Pepper, R.J. Swart, J.A. Edmonds, L.G. Meira Filho, I. Mintzer, M.X. Wang and J. Watson, "Emissions scenarios for the IPCC, an update: assumptions, methodology, and results," in Climate Change 1992: The Supplementary Report to the IPCC Scientific Assessment, eds. J.T. Houghton, B.A. Callander and S.K. Varney (Cambridge: Cambridge University Press 1992).

[xxiv] Belize, "First National Communication to the Conference of the Parties of the United Nations Framework Convention on Climate Change" (July 2002) http://unfccc.int/national_reports/non-annex_i_natcom/items/2979.php

[xxv] Cuba, "First National Communication to the Conference of the Parties of the United Nations Framework Convention on Climate Change" (September 2001) *http://unfccc.int/national_reports/nonannex_i_natcom/items/2979.php*

[xxvi] Dominican Republic, "First National Communication to the Conference of the Parties of the United Nations Framework Convention on Climate Change" (June 2003) http://unfccc.int/national_reports/nonannex_i_natcom/items/2979.php

[xxvii] Guatemala, "First National Communication to the Conference of the Parties of the United Nations Framework Convention on Climate Change" (December 2001) http://unfccc.int/national_reports/nonannex_i_natcom/items/2979.php

[xxviii] Based on Ortiz (2007a), Ortiz (2007b), Ortiz et al (1998) and Ortiz and SOPASAL (2000)

[xxix] Republique D'Haiti, "Premiere Communication Nationale sur les Changements Climatiques," (August 2001) http://unfccc.int/national_reports/non-annex_i_natcom/items/2979.php.

[xxx] Ministère de L'Environnement, Programme Changements Climatiques, République d'Haïti, Plan D 'Action National D 'Adaptation (October 2006).

[xxxi] United Nations Framework Convention on Climate Change http://unfccc.int/essential_background/library/items/3599.php?rec=j&priref=5838 (accessed 21 May 2009).

[xxxii] Honduras, "First National Communication to the Conference of the Parties of the United Nations Framework Convention on Climate Change" (November 2000) http://unfccc.int/national_reports/nonannex_i_natcom/items/2979.php

[xxxiii] Mexico, "Third National Communication to the Conference of the Parties of the United Nations Framework Convention on Climate Change" (November 2006) http://unfccc.int/national_reports/nonannex_i_natcom/items/2979.php

[xxxiv] C. Conde, R. Ferrer and S. Orozco, "Climate change and climate variability impcts on rainfed agricultural activities and possible adaptation measures: a Mexican case study," *Atmósfera 19* (2006), no. 3: 181-194.

[xxxv] C. Conde, R. Ferrer and S. Orozco, "Climate change and climate variability impcts on rainfed agricultural activities and possible adaptation measures: a Mexican case study," *Atmósfera 19* (2006), no. 3: 181-194, p. 183.

[xxxvi] M.B. Wehbe, R.A. Seiler, M.R. Vinocur, H. Eakin, C. Santos and H.M. Civitaresi, "Social methods for assessing agricultural producers' vulnerability to climate variability and change based on the notion of sustainability," AIACC Working Paper 19, Assessment of Impacts and Adaptations of Climate Change Project (funded by the Global Environment Facility,

US Agency for International Development, the Canadian International Development Agency and the US Environmental Protection Agency) (2005), www.aiaccproject.org.

[xxxvii] C.A. Salinas-Zavala and D.B. Lluch-Cota, "Relationship between ENSO and winter-wheat yields in Sonora, Mexico," *Geofísica Internacional 42* (2003), no. 3: 341-350.

[xxxviii] A.L. Luers, D.B. Lobell, L.S. Sklar, C.L. Addams and P.A. Matson, "A method for quantifying vulnerability, applied to the agricultural system of the Yaqui Valley, Mexico," *Global Environmental Change 13* (2003): 255-267.

[xxxix] A.L. Luers, D.B. Lobell, L.S. Sklar, C.L. Addams and P.A. Matson, "A method for quantifying vulnerability, applied to the agricultural system of the Yaqui Valley, Mexico," *Global Environmental Change 13* (2003): 255-267, p. 265.

[xl] R. Boyd and M.E. Ibarrarán, "Extreme climate events and adaptation: an exploratory analysis of drought in Mexico", *Environment and Development Economics* (in press).

[xli] E. Umaña, "Agroambiente y Cambio Climatico. Impactos del Cambio Climatico Sobre Sistemas Agrarios Medio Ambiente" (Nicaragua: Universidad Nacional Agraria, Facultad De Recursos Naturales y Der Ambiente [FARENA] 2003).

[xlii] D. Espinosa, A. Mendez, I. Madrid and R. Rivera, "Assessment of climate change impacts on the water resources of Panama: the case of the La Villa, Chiriqui and Chagres river basins, *Climate Research 9 (*1997): 131-137.

[xliii] N. Guindin, A. Weiss and L. Pérez-Alegría, "Impact of climate change on water availability for a bioenergy project in the Lajas Valley, Puerto Rico.

[xliv] Intergovernmental Panel on Climate Change (IPCC), "Glossary," in *Climate Change 2007: Impacts, Adaptation and Vulnerability: Contribution of Working Group* II to the Fourth Assessment Report of the Intergovernmental Panel on Climate Change, eds. M.L. Parry, O.F. Canziani, J.P. Palutikof, P.J. van der Linden and C.E. Hanson (Cambridge: Cambridge University Press 2007).

[xlv] G. Yohe and R. Tol, "Indicators for social and economic coping capacity—moving toward a working definition of adaptive capacity," Global Environmental Change 12 (2002): 25-40.

[xlvi] E.L. Malone and A.L. Brenkert, "Vulnerability, sensitivity, and coping/adaptive capacity worldwide," in The Distributional Effects of Climate Change: Social and Economic Implications, eds. M. Ruth and M. Ibarraran (Dordrecht: Elsevier Science in press).

[xlvii] UNEP-GRID-Arendal, "Climate impacts of El Niño phenomenon in Latin America and the Caribbean, Maps and Graphics Library (2005)
http://maps.grida.no/go/graphic/climate (Accessed 18 May 2009).

[xlviii] M.L. Parry, O.F. Canziani, J.P. Palutikof, P.J. van der Linden and C.E. Hanson, eds., *Climate Change 2007: Impacts, Adaptation and Vulnerability: Contribution of Working Group II to the Fourth Assessment Report of the Intergovernmental Panel on Climate Change (Cambridge:* Cambridge University Press 2007).

In: Implications of Climate Change in Mexico... ISBN: 978-1-61122-849-6
Editors: Samuel B. Duncan

Chapter 2

MEXICO, THE CARIBBEAN, AND CENTRAL AMERICA: THE IMPACT OF CLIMATE CHANGE TO 2030: GEOPOLITICAL IMPLICATIONS

National Intelligence Council

SCOPE NOTE

Following the publication in 2008 of the National Intelligence Assessment on the National Security Implications of Global Climate Change to 2030, the National Intelligence Council (NIC) embarked on a research effort to explore in greater detail the national security implications of climate change in six countries/regions of the world: India, China, Russia, North Africa, Mexico and the Caribbean, and Southeast Asia and the Pacific Island States. For each country/region we are adopting a three-phase approach.

- In the first phase, contracted research explores the latest scientific findings on the impact of climate change in the specific region/country. For Mexico, Central America, and the Caribbean, the Phase I effort was published as a NIC Special Report: **Mexico, Central America, and the Caribbean: Impact of Climate Change**

to 2030, A Commissioned Research Report (NIC 2009-11), of November 2009.

- In the second phase, a workshop or conference composed of experts from outside the Intelligence Community (IC) determines if anticipated changes from the effects of climate change will force inter- and intra-state migrations, cause economic hardship, or result in increased social tensions or state instability within the country/region. This report is the result of the Phase II effort for Mexico, Central America, and the Caribbean States.
- In the final phase, the NIC Long-Range Analysis Unit (LRAU) will lead an IC effort to identify and summarize for the policy community the anticipated impact on US national security.

In August of 2009, a group of regional experts convened to explore the socio-political challenges, civil and key interest group responses, government responses, and regional and geopolitical implications of climate change on Mexico, Central America, and the Caribbean through 2030. The group of outside experts consisted of economists, political scientists and other social scientists. While the targeted time frame of the analysis was out to 2030, the perceptions of decision makers in 2030 will be colored by expectations about the relative severity of climate changes projected later in the century. The participants accordingly considered climate impacts beyond 2030 where appropriate.

This work is being delivered under the Global Climate Change Research Program contract with the CIA's Office of the Chief Scientist.

Executive Summary

The National Intelligence Council-sponsored workshop entitled *The Implications of Global Climate Change in Mexico, the Caribbean, and Central America*, held on 31 August 2009, brought together a panel of regional experts to consider the probable effects of climate change on Mexico, the Caribbean, and Central America from a social, political, and economic perspective. The panelists concluded that through 2030 *climatic changes in the region may aggravate civil unrest and internal conflicts leading to increased migration, and that strong, centralized states, and states with robust civil societies, will likely fare better than others.*

Although the region does not contribute to significant global greenhouse gases, it is highly vulnerable to the effects generated by increasing climate variability. Rising temperatures, rising sea levels, increased rainfall in some places, drought in others, and a greater frequency of extreme weather events such as hurricanes, floods, and heat waves are expected from climate change.

- Temperature rise, both on land and at sea, could affect crop growing patterns as well as the viability of fisheries on which some coastal populations depend.
- Sea level rise has the potential to create economic losses for coastal populations, particularly around ports.
- Rising temperatures, increased rainfall, and population displacement may exacerbate or contribute to the emergence of infectious diseases.

Climate change may increase prospects for conflict in the region. Increased resource scarcity may combine synergistically with weak states and economic inequalities to promote organized insurrection. Civil conflicts are more likely than state-on-state violence, though some internal conflicts may spill across borders.

- Potential inter-state flashpoints include the Dominican Republic and Haiti, the GuatemalaHonduras-Nicaragua zone of instability, and the US-Mexican border.
- In Mexico, water scarcity may intensify community vulnerabilities to water-related diseases and exacerbate social tensions over access to water for agriculture and domestic uses.

Large internal conflicts or large population migrations are unlikely for most states in the region (with the exception of Haiti). While increased migration is likely, it is only part of a range of potential responses, including adaptation, sub-state conflict, and interstate conflict.

- Migratory trends to Mexico and to the United States are likely to continue and may accelerate. A concomitant rise in migrant-related criminality is also likely.
- The adaptive capacities of states and populations within the region vary but the region has historically dealt well with slow-acting climatic changes.

Strong civil society organizations play an important role in moderating the effects of climate change as they provide a means of communication between local people and state actors. The strength of civil society varies among states in the region.

- Mexico and Panama possess robust civil society organizations, environmental and social NGOs, and ethnic organizations.
- Guatemala, Nicaragua, and El Salvador are in the process of rebuilding civil society destroyed by conflict.
- In the case of Haiti, civil society is more present in Diaspora—primarily in the United States and France.

State centralization and control correlate well with emergency preparedness. All states in the region show some capacity to recover from damage caused by extreme weather events.

- Mexico and Cuba appear to have the greatest state-level capacity when it comes to emergency preparedness. In both settings, the armed forces play a role in managing emergency response.
- The smaller Caribbean and Central American states are less prepared to respond to climatic emergencies domestically, and Haiti is the least prepared.

Most states in the region lack the institutional mechanisms to effectively address the long-term threats posed by climate change.

- Development agencies including USAID and the Inter-American Development Bank are working with states in the region to gather information and develop tool kits to help governments assess their vulnerabilities and better prepare for the physical, economic, and human toll of extreme weather events.
- The Caribbean Community (CARICOM) has instituted several initiatives to bring governments and NGOs together in productive cooperation.

The United States will probably face ever-increasing pressure to provide humanitarian assistance to neighbors to avoid large numbers of refugees and to reduce the risk of local conflicts that could require US military intervention.

- States in the region may become increasingly inward looking and politically populist in reaction to economic disparity, resource scarcity, or civil conflicts exacerbated by climate change.
- Seeking collaborative and effective regional mechanisms to jointly manage climate change challenges may provide the United States new opportunities for strengthening relations with countries in the region.

INTRODUCTION AND BACKGROUND[1]

Mexico, the countries of the Caribbean, and Central America examined in the Phase I report are at risk from the impacts of climate change in the next 20 years because they will be exposed to a greater range of climate changes and have a relatively weak adaptive capacity when compared to the world at large. Within the region, climate change is evident in increased temperatures, changes in precipitation, and sea level rise—and perhaps in weather variability and natural disaster events. Countries considered for the Phase I effort included Belize, Cuba, the Dominican Republic, Guatemala, Haiti, Honduras, Mexico, Nicaragua, and Panama; Puerto Rico was also discussed.

Steady increases within the region in the number of extreme weather events—hurricanes, storms, and droughts—and their effect on infrastructure, public health, loss of human life and agriculture may be attributable to climate change. The countries reviewed do not yet have a full understanding of the potential impacts of future climatic changes and are not prepared to prevent or reduce those impacts.

Regional leaders are aware of these challenges and have begun to make commitments and agreements that will enhance their understanding of future climate change, their own adaptive capacity, and where critical changes and investments need to be made. Leaders have not addressed the problem from a preventive perspective through policy changes or infrastructure investments because of a lack of systematic analysis that quantifies and qualifies the potential impact to the region, allowing the development of relevant and economically viable options. At present the region is still responding to climate change in a reactive manner.

- Regional leaders realize that leaving the situation "as is" will exacerbate their fragile economies, resources, and adaptive capacity but lack strategic plans to address the issue.

- Most countries in the region are signatories to many multilateral environmental agreements but are only now beginning to implement such agreements.
- There are significant gaps in the ability to fully understand in a systemic way all the dimensions of climate change impacts at the economic, social, and/or environmental level in the region. There are gaps and deficiencies in data, systematic methodologies/analysis, and tools to monitor, share, and track information and events at the local, national, and regional levels.

Efforts are starting to reduce systemic knowledge gaps. There is insufficient funding by regional governments to undertake detailed modeling that would result in information to rank and evaluate the financial viability of potential climate change adaptation projects. Several entities at the national and regional levels are working to develop improved analytical methods and information sharing as well as better data and data availability.

- In September 2008, the Economic Commission for Latin America and the Caribbean (ECLAC) announced that it would undertake multiple studies to review how climate change is affecting regional economies. Currently, the ECLAC consensus is that climate change is likely to impose serious economic consequences for the Central American and Caribbean regions, making it increasingly difficult to respond to the challenges of poverty reduction, higher human development, and environmental sustainability linked to the attainment of the United Nations Millennium Development Goals.
- Upcoming studies by the ECLAC are expected to contribute to a better understanding of the economic impact of climate change in the region and will outline the costs and benefits of needed related policy responses, both in terms of mitigation and adaptation.

In the Phase I report, information available for a selected set of Mexico, Caribbean, and Central American countries was reviewed to start understanding the projected climate change variability, given certain scenarios to 2030, as well as to start an initial assessment of these countries' current adaptive capacity to reduce such effects.

Very limited modeling and analysis are available for the countries of interest. Because of that, this initial analysis draws heavily on the respective governments' First National Communication to the Conference of the Parties of the United Nations Framework Convention on Climate Change (UNFCCC). These reports to the UNFCCC offer the most comprehensive and comparable information available today. In the case of Mexico, the third communication to the UNFCCC was used to review summary impacts. This report, however, was primarily focused on improving inventories of greenhouse gases across all types and production of energy as well as the greenhouse gases generated by major economic activity.

This review identifies the following high-priority risks:

- *Energy*. Energy resources, production, and use vary widely across the countries under review. As all the countries experience population growth, economic growth, and industrialization, they will increase their need and demand for energy. All the countries under review rely on imported fossil fuels, with the exception of Mexico, which is a net exporter of energy resources. In most of the countries, the largest generator of greenhouse gases is the energy sector. Although they are very small contributors to global emissions, most of the countries will receive economic benefits from increasing use of renewable energy. Most have begun efforts to evaluate and implement small renewable energy projects, such as solar energy in rural areas of El Salvador, wind energy in Nicaragua and Costa Rica, and an intensive effort in the Dominican Republic to evaluate hydro-generated electricity.
- *Agriculture*. The agricultural sector climate-related research for most of the countries in this review is limited. Where research is available, productivity losses are projected for optimistic, moderate, and pessimistic scenarios for some key food crops with estimates that vary from 10 percent to more than 50 percent degradation by the year 2030.
- *Water Resources*. The majority of the population in most of the countries in the review lives in coastal areas, which are highly vulnerable to severe climate changes. As populations continue to grow in the same areas, increasing water extraction and rising sea levels are expected to have severe impacts on the quantity and quality of water available. Many of these countries' aquifers are open to ocean waters and are already experiencing increased salinity. Rising sea levels will accelerate the deterioration of aquifers and available water resources.

- *Migration.* In Central America, an increase in intra-regional migration during the 1 980s and 1990s as well as extra-regional migration was the result of social unrest and economic contraction. Future patterns of migration are not expected to change significantly.

Moreover, the inability of countries in the region to adapt and recover from severe climate events with major impacts on their economies will continue to promote migration outside the region, in particular to the United States and Canada. The large number of immigrants coming to the United States in the past 20-25 years will facilitate this movement.

Most of the countries under review have submitted their First Communication to the UNFCCC; Mexico has submitted its third. Significant work and analysis needs to be done to fully capture the impact on socio-economic systems and their current ability to recover, adapt, and reduce the effects of climate change.

The great variation of information available for each country reduces the ability to compare the full set of key indicators across all countries in a consistent manner.

Social, Political, and Economic Challenges

The projected primary effects of climate change in Mexico, Central America and the Caribbean include rising temperatures, rising sea levels, increased rainfall in some places, drought in others, and a greater frequency of extreme weather events such as hurricanes, floods, and heat waves. Projected secondary effects include the emergence or re-emergence of infectious diseases, shifting geographic ranges of crops for consumption and export, saltwater contamination of aquifers, and changes to animal habitats and biodiversity, among others.

In the short term, the extent to which these factors may exacerbate social tensions regarding control over and use of natural resources, provoke population movement within and from states, contribute to food scarcity, or lead to pandemic outbreaks will depend on the ability of governments and civil society groups in the affected states to prepare for and respond to emergencies, adapt to geographic and social changes, and develop ways to cooperate and communicate with international partners in emergency situations. In the long term, strengthening policies to contain states' contributions to further global warming through the emission of green house gases and deforestation will be essential.

Meteorological Challenges

Climate experts project that extreme weather events due to higher ocean water and air temperatures, including more frequent hurricanes, increased rainfall, droughts, and heat waves will characterize the effects of climate change in Mexico, Central America, and the Caribbean. Beyond loss of life and economic impacts, extreme weather events can displace people and create the conditions for disease outbreaks. In the Americas, the states that border the Caribbean and Gulf of Mexico bear the brunt of hurricane action.

- In 2008, Hurricane Ike devastated parts of Cuba, Haiti, and Jamaica. Heavy rains in the eastern Mexican state of Tabasco in 2007 covered at least 80 percent of the state with water, provoked a national emergency declaration, and dislocated at least 20,000 people.
- In 2007, Hurricane Felix, a category 5 storm, made landfall on the Nicaraguan Coast. Storms of this intensity could be the norm for future storms.
- In 1998, Hurricane Mitch, the second worst Atlantic hurricane on record, not only destroyed roads and bridges in Honduras, Nicaragua, and Costa Rica but also wreaked havoc in the Caribbean and Mexico. At least 11,000 people are believed to have died, and Mitch is estimated to have caused six billion dollars in damage. Outbreaks of cholera, acute respiratory infections, and dengue fever were also reported.

Mexico and Cuba appear to have the greatest state-level capacity when it comes to emergency preparedness. When Hurricane Mitch hit Central America in 1998, both states responded by sending personnel to help with rescue and reconstruction efforts. In both settings, the armed forces played a role in managing emergency response. Mexico has also engaged in emergency preparedness exercises with the United States and Canada through the former Security and Prosperity Partnership and in the context of the Global Health Security Action Group. However, the smaller Caribbean and Central American states seem to be less prepared to respond to emergencies and extreme weather events domestically; for example, Haiti is heavily reliant on external support for hurricane and flood recovery.

Agricultural Challenges

Mexico, Central American, and Caribbean states have economies with significant agricultural sectors. In many of the states selected for this review, the agricultural land use vis-à-vis total land area varies widely. In 2005, Belize agricultural land was 6 percent of the total land area, reflecting the fact that over 50 percent of GDP comes from the services industry, particularly tourism, compared to Dominican Republic where agricultural land was 70 percent, Costa Rica and Haiti 57 percent, Cuba 60 percent and Mexico 55 percent of total land area. In the past 27 years, all of the states reviewed have maintained relatively stable ratios of agricultural land use to total land area.

The conversion of forests to agricultural use is likely to continue; however, the general projected drying trend in the area is likely to limit the types of viable agricultural crops. Although projected temperature changes may not differ much by season, changes in rainfall will likely lead to extended periods of drought and possible loss of soil fertility during the peak growing season in June, July, and August.

- Projections for productivity losses in Cuba range from 10 to 15 percent for rice, cassava, and corn; five to 10 percent for sugar cane; and 40 to 45 percent for potatoes.
- Coffee production in Veracruz, Mexico is likely to drop over 30 percent by 2020, degrading its economic value to the region.

Projections for other states in the region are similar, though they vary by state and study. Further, the rise in CO_2 levels could result in a fertilizing effect with crops having shorter growing cycles. The salinization of ground water supplies due to climate change and sea-level rise may also threaten agricultural productivity.

Many Central American and Caribbean states have major fishing industries. Climate change is likely to lead to changes in migration patterns and depth of fish stocks with possible negative effects on the fishing industry.

Mexico, the Central American states, and the Dominican Republic are engaged in free trade agreements with the United States, and some of the Caribbean states have established trade pacts with the European Union. These agreements regarding the flow of goods, including agricultural products and foodstuffs, may reduce some populations' reliance on locally grown crops. However, climate change is likely to negatively affect the fortunes of subsistence farmers, such as those in Mexico, Haiti, Guatemala, who are not engaged in commercial agricultural activities.

Coastal and Maritime Challenges

Climate change scenarios project that sea levels will rise, a phenomenon which could be a significant factor for the Caribbean islands and for coastal communities in Mexico and Central America. Sea level rise has the potential to create economic losses for coastal populations, particularly ports that could not only experience population dislocations but also shifts in commercial traffic.

The small islands of the Caribbean will probably experience a warming over the next century that may be somewhat smaller than the global annual mean warming. Projections for temperature increases in the Caribbean at the end of the 21st century range from 1.4 °C to 3.2 °C with a median of 2.0 °C. This level of warming is still likely to lead to significant sea level rise, deterioration of coastal areas and erosion of beaches, and increased invasion of non-native species, while reduced water resources could lead to an inadequacy of fresh water to meet demand during low rainfall periods. The amount of sea level rise is not likely to be uniform due to geographical differences in the islands. Extensive geographical, topographical, ecological, sociological, and population density information, gathered into a detailed geographic information system (GIS), is required before predictions are possible.

A rise in sea levels could also aggregate social tensions in many of the regional states. European or mestizo settlers have traditionally pushed indigenous populations in Mexico and Central America away from the most productive land to coastal areas or mountainous zones. Were sea level rise to provoke population movement, settled indigenous populations could experience further pressure from dislocated populations to move toward even less productive soil. This could further intensify emigration from marginal communities, creating conditions in which men migrate to more lucrative areas to seek access to land or income, while women and children bear the responsibility of maintaining the household and become vulnerable to further land-grabbing efforts.

Having suffered centuries of racial discrimination, often at the hands of state officials, socially marginal communities, including indigenous populations and those of African descent, may be cautious about official efforts to relocate them or mediate conflicts among communities. This is certainly the case in Mexico and Guatemala and most likely holds true for some Caribbean communities, as well. Collective action focused on the needs of indigenous peoples has characterized political movements in both states, although the extent to which such movements might gather greater support in the future is unclear.

Hydrologic Challenges

Depletion of aquifers and reduced rainfall in some areas will contribute to water scarcity, which may intensify community vulnerabilities to water-related diseases and exacerbate social tensions over access to water for agriculture and domestic uses. In Mexico, conflicts over water use have created tensions in local communities throughout the state, provoking marches, blockades, and efforts to take over institutions.

Mexico City is experiencing severe water scarcity and aquifer depletion. The city already sits on drained lakes, exacerbating infrastructure vulnerable to seismic activity. With a population of more than 20 million, the city must pump water from great distances and has had to ration water at least three times in 2009. Informal urban settlements rarely count on regular water service, and residents of such neighborhoods may have even more restricted access to water with rationing in effect. Irregular supply limits users' access to water for hygienic purposes and can lead to contamination when pipes are not flushed out on a regular basis.

Water scarcity can lead to tensions between states as well. It has already created conflicts between Mexico and the United States over the Rio Grande, as well as Mexico and Guatemala over the Usumacinta River. Most of the states in Central America share some form of water boundary, suggesting that measures must be put in place to resolve conflicts at state and community levels.

Demographic and Public Health Challenges

Mexico, Central America, and the Caribbean states all continue to experience population growth, albeit at somewhat different rates, leading to an increase in food demand. Most of the states in these regions depend greatly on agricultural production. Variations in crop yields, food crops, and cash crops present major challenges.

Increased rainfall, sea level rise, drought, and extreme weather events may provoke populations to migrate to more suitable habitats. The arrival of environmental migrants to existing settlements may provoke tensions with local populations and competition over scarce resources. Migration may also lead to the separation of families, with males leaving home to seek income-generating opportunities, placing the burden of household maintenance on women and older children.

Rising temperatures, increased rainfall, and the movement of populations into new areas may exacerbate or contribute to the emergence (or re-emergence) of infectious diseases, including diarrheal disease and acute respiratory disease, as well as vector-borne diseases such as dengue fever, malaria, leptospirosis, and Chagas disease.

- Since 1990 the region has experienced a series of re-emerging diseases following severe climatic events such as floods, hurricanes, and droughts.
- There is evidence of increases in several communicable diseases, such as dengue, malaria, and Hantavirus pulmonary syndrome; and the reemergence of a large host of infectious diseases following years in which there were El Niño/Southern Oscillation (ENSO) events.

While Mexico, Costa Rica, and Cuba have relatively strong health systems, many of the Central American and Caribbean states' health systems are weaker. The migration of health professionals from Mexico, the Caribbean, and Central America to the United States and Canada, and the fact that so many people do not have access to health care, creates vulnerabilities. Health systems in the region are already burdened by the increasing toll of chronic disease and will need to develop tools to anticipate climate-related outbreaks and develop lab capacity and response to address problems.

Economic and Energy Challenges

Since 1990, the region has experienced large disparities in states' GDPs. Some low values have been the result of economic contraction coupled with political unrest, capital flight, migration of the better-educated segment of the population, and the loss of foreign investments as experienced by Guatemala, El Salvador, Nicaragua, and Haiti from the late 1970s through the 1990s. The socio-political challenges of the 1 980s and increases in extreme weather events in the 1 990s had adverse effects on the fragile economies of the region. Instability in the economies exacerbated by the absence of law has greatly reduced the opportunity for recovery. States such as El Salvador, Guatemala, and Nicaragua were directly affected by civil unrest and increased weather-related natural disasters, while their neighboring states had to cope with an increase in migrating population because of the difficulties associated with war and natural disasters. These same states have also been severely affected by hurricanes, floods, and tropical storms in the past two decades.

The potential for resource scarcity, especially water, to negatively affect the economies in the region is highly probable. Many states have agriculture-based economies whose potential negative impact can be assumed. However, states with growing manufacturing sectors—

Mexico, El Salvador, Costa Rica, and Guatemala—also face challenges from scarcity in addition to increased competition from China and other regions. Tourism may also decrease along with remittances from immigrant populations abroad as those regions face their own economic troubles.

The states in the region, which have mostly fossil fuel-based economies, are mostly net importers of sources for energy production. Since 1984 they have continued to increase their overall energy consumption. With the exception of Mexico, primary and secondary energy production has remained below total annual consumption. Primary energy production refers to the production of energy products or sources found in their natural states, such as wood, natural gas, bagasse, and hydroelectricity. It also includes the amount of fuel extracted and the energy consumed in the production process and the supply to energy producers and conversion. Secondary energy production includes products or sources that are the result of conversion of primary energy products such as all those derived from petroleum refining, kerosene, and diesel.

As economies grow and the process of industrialization continues, most states in the region will remain highly vulnerable to variable petroleum base supply and cost as experienced in the past few years. Mexico is the only state in the region that is a net exporter of energy resources, though this is likely to change within the next decade. Between 1990 and 2007, regional energy consumption increased 158 percent. Moreover, Costa Rica, Nicaragua, and the Dominican Republic increased consumption by about 200 percent while Panama increased by 288 percent in the same period. Energy consumption is expected to increase as population and economies continue to grow.

Energy supply composition across the states remains predominantly based on petroleum, with the exceptions of Haiti, Nicaragua, and Honduras. Interestingly, these are the three states with the lowest annual GDP growth rates within the group from 1990 to 2007. On the other hand, Costa Rica, Cuba, Panama, and the Dominican Republic, who experienced the largest annual GDP growth rates in the same period, also have the largest shares of oil-based energy supply.

With the exception of Mexico, all states are net importers of petroleum-based products. Oil- based energy supply remains significant in particular in the case of the Dominican Republic, where it accounted for 74 percent of total energy supply in 2005 and 79 percent in 2002. Island nations such as Cuba, Puerto Rico, Haiti, and the Dominican Republic remain particularly vulnerable to the supply of petroleum-based energy products, since the raw material must be brought to the islands by ship for refining and processing. The recent discovery of oil reserves off the coast of Cuba should provide some improvement, but to what extent remains unclear. Hydroenergy plays a significant role only in Costa Rica, where it accounts for 18 to 24 percent of supply; for the rest of the states it ranged from 0.1 percent in Cuba to 9.8 percent for Panama.

Civil and Key Interest Group Responses

Potential Responses

Evidence from previous post-disaster research challenges the idea that full-scaled "flight-orfight" responses are likely for populations in most states in the region, with the notable exception of Haiti, at least as a direct response to the impacts of climate change alone. Rather than limited to "flight-or-fight," there are a range of responses to climate change-induced hazards that may or may not be adapted at varying spatial, organizational, and temporal scales.

Innovation/Intensification/Adaptation. Historically, people and societies have been able to respond to natural hazards and human-generated alterations in the socio-ecological system by adopting innovative (innovation/adaptation) and non-innovative (intensification) changes to ensure economic survival. Adjustments to living and livelihood patterns can stress people and governments but probably significantly less than the alternatives that follow below. Adjustments will be geared toward relevant stressors in the different states under consideration and can range from wholesale agricultural alterations in the crop or techniques to improve construction of buildings to withstand extreme climatic conditions. Hurricanes and other storms demand technological adjustments, while slower-acting change processes such as warming or changes in precipitation may respond well to modifications of existing production techniques. These adjustments are most likely the least socially, economically, and politically disruptive responses to climate change and will be aided by strong civil society institutions due to their ability to moderate relationships between localities and regions and central governments.

Abandonment/Migration. Many people in the region currently use temporary or permanent migration as a coping mechanism for socio-economic stressors brought on by economic or political hardship and occasionally by climate related phenomena. Large groups of people have previously relocated due to changes in resource availability, among other reasons. In worst-case scenarios, large-scale migration might occur out of areas and perhaps across international borders. This migration could place more pressure on an already stressed system and potentially lead to internal or international conflict. There are, however, many moderate migratory activities that can take place—such as rural to urban and rural to rural—all within a single country. States in the region have been urbanizing rapidly since the end of the Second World War, as has international migration. Slow-acting climate change processes may signal almost imperceptible increases in movement while faster acting change can challenge socio-ecological resilience.

New migrants, often representing the extremely poor, may pose a threat to the existing social order and potentially give rise to violent backlash. Criminal elements such as narcotic traffickers and smugglers often accompany migrants, exacerbating existing tensions between and within states. Recent examples of migration in the region include Haitians to the Dominican Republic; Guatemalans to Mexico and Belize; Salvadorans, Mexicans, Haitians, and Dominicans to the United States; and Nicaraguans and Panamanians to Costa Rica. Except for the 1969 Soccer War between El Salvador and Honduras, these movements resulted in muted violence, usually at the individual or occasionally community level.

Sub-state Conflict. The states in the region have varying experience with internal conflict throughout their histories. Four states—Guatemala, Nicaragua, Haiti, and Mexico—have had or currently have significant conflict-related, political ideologies fueled by drastic economic inequalities, while Honduras, Belize, and the Dominican Republic have remained relatively peaceful over the last 25 years. None of these conflicts can claim climate change as a necessary or sufficient condition for conflict. Increased resource scarcity, however, may combine synergistically with weak states and economic inequalities leading to organized insurrection. Groups already organized for conflict (though not resource conflict *per se*) include Mexico's Zapatistas (other similar movements exist across Mexico), Mexico's drug cartels, and loosely organized gangs like MS-13 in Central America.

Interstate Conflict. There exist precedents for cases of interstate conflict related to resource issues, among other reasons, and there is the potential for more to occur especially as migration and intra-state conflict stress developing-state governments. In the region of Mexico, Central America, and the Caribbean, four major drivers of resource scarcity could lead to increased interstate violence: Agricultural land loss or degradation, forest loss and degradation, fresh water depletion or pollution, and fisheries depletion.

Adaptive Capacity

Most estimates of vulnerability rely on easily attainable statistics—economic capacity, human capital, and environmental capacity—that ignore less quantifiable but important factors on how human-natural systems interrelate, namely social interaction. Such estimates fail to recognize the enormous adaptive capacity most societies possess under the right social conditions.

At the local level, adaptation may result from individual or group innovation but appears unlikely to be widespread without governmental and civil society cooperation. Civil society in the region has a long, if uneven history. States like Mexico possess robust civil society organizations such as the Catholic Church, environmental and social non-governmental organizations (NGOs), and ethnic organizations. Other states such as Guatemala are in the process of rebuilding civil society, much of which was targeted during that state's civil war. In some cases, such as Haiti, civil society may be more present in migration destinations— primarily the United States and France—due to internal problems with Haitian social and economic interactions.

It is in enhancing adaptive capacity that civil society may play its most important role. In most of the region, important state functions such as environmental management, food distribution, and health care have been ceded to civil society of a religious, environmental, or ethnic nature.

- The Mexican NGO PRONATURA, or international NGOs, manage most parks and reserves in Mexico due to the state's inability to pay for and monitor environmental conditions.
- Haitian food distribution networks are largely based on the role of the Catholic clergy and laity or the increasing number of non-Catholic, Christian organizations.
- Domestic and international NGOs in Guatemala communicate ethnic demands and concerns to the state.

Development agencies including USAID and the Inter-American Development Bank are working with states in the region to gather information and develop tool kits to help governments assess their vulnerabilities and better prepare for the physical, economic, and human toll of extreme weather events. Additionally, the Caribbean Community (CARICOM) has instituted several initiatives to bring governments and NGOs together in productive cooperation.

- The Instituto de Nutrición de Centroamérica y Panamá (INCAP), based in Guatemala, provides advice and technical support to governments working to resolve nutrition problems associated with food scarcity. Under nutrition continues to be a challenge for some populations in Mexico, Central America, and the Caribbean, even as problems related to obesity, diabetes, and cardiovascular disease are beginning to pose economic and social challenges to regional health systems.
- In the Caribbean, CARICOM representatives agreed to establish the Caribbean Public Health Agency (CARPHA) to improve disease surveillance in the region.
- Belize hosts the CARICOM Caribbean Community Climate Change Research Center, which serves as a regional repository for climate data and conducts research on the projected effects of global warming in the region. The Center also provides climate change data to member states to help them prepare for the United Nations Framework Convention on Climate Change (UNFCC) Conference of Parties in Copenhagen in December of 2009. Given that the states in the region do not significantly contribute to global warming, the Center provides considerable economic data charting the effects of warming on states' economies and the projected effects.
- The Caribbean Environmental Health Institute (CEHI), a CARICOM entity based in St. Lucia, provides technical assistance on water safety plans and on water-related climate planning throughout the region.
- A partnership involving the Pan American Health Organization, the International Water Association, and the Asociación Inter-Americano de Ingenieros Sanitarios (AIDIS) is introducing states to water safety to facilitate water quality planning at the catchment level, with experts from Honduras, Mexico, and Jamaica playing key roles in the organization.

In addition, the goals of many civil society organizations focus on capacity building at the local level. Knowledge dissemination (education) and production (research) carried out locally promises to enhance adaptive capacity through increased recognition of problems associated with climate change, alternative economic activities, and ability to communicate with the government when problems arise. This capacity building can be aided by international cooperation with international NGOs, government agencies, and private enterprise from the United States and elsewhere.

In those states with high levels of political pluralism, civil liberties, and participation (Panama, Mexico), civil society has the potential to play a positive role in mitigating the impacts of climate change. Weaker democracies, such as Guatemala, Nicaragua, and El Salvador, or Honduras, continue to strive to recover from civil war or natural disaster. It is difficult to predict how strongly centralized states (Cuba) or chronically disorganized states (Haiti) will be able to adapt to the predicted changes associated with climate change. Should adaptive capacities fail to meet the needs of climatic challenges, large-scale flight is likely with the United States and Mexico bearing the largest immigration burdens. In the case of Haiti, this is a near certainty.

State Responses

The region of Mexico, Central America, and the Caribbean lags most other major regions as an emitter of greenhouse gases that are believed to contribute to global warming, yet it is on the frontlines of the potential consequences, including mega-storms, droughts, and impacts on agriculture, eco-systems, and epidemiology. This blunt disparity is even greater for the small and micro-economies that make up most of Central America and the Caribbean. The region's relative paralysis and reactive approach to climate change may be driven primarily by other demands for their scarce resources, and a recognition that limiting carbon emissions within the region will barely impact the overall trajectory of global climate change.

Any effective effort to combat the host of climatic threats to the region must overcome other challenges in the region to include weak institutions, inadequate or deteriorating infrastructure, pollution, unmanaged deforestation, high levels of poverty, overcrowding of urban areas, and inefficient means of agricultural production. If fishermen and subsistence farmers are unable to earn a living through traditional activities because of global warming, they will need to learn new ways of generating income to support their families. It will be important for governments and development agencies to help such populations identify alternative livelihoods.

International Efforts

It is currently unclear how effective regional and international bodies can be, though several exist in many of the states in question. In many cases, state governments have partnered with international NGOs as a means of interstate cooperation and coordination.

- Health Ministers cooperate through the Sistema de Integración de Centroamérica (SICA) and CARICOM, and health is a recurrent theme at the Summits of the Americas.
- For the Caribbean states, participation in the United Nations Small Island Developing States Network has enabled some governments to gather information about the projected effects of sea level rise and to anticipate the changes that will affect their territories. Through their involvement in CARICOM processes, governments in the Caribbean participate in regional discussions about climate change preparedness and adaptation.
- Mexico, along with Canada and the United States, is a member of the Global Health Security Action Group (GHSAG) and has worked with Canada and the United States to develop mutual assistance protocols in the event of a public health emergency.

States in the region have also undertaken their own initiatives alone and in conjunction with neighbors.

- In 2008, Mexico's President Felipe Calderón proposed the establishment of a global "Green Fund" to provide support to states seeking to develop their capacities to adapt to climate change scenarios.
- In 2004, Mexico hosted the fourth World Water Forum, making a special effort to highlight water issues in the region.
- Mexico and Guatemala established an International Commission on Limits and Waters in 1961 that may serve as an example to others.

Mexico and Cuba have the greatest potential to play a global leadership role in combating climate change, although the small-island states of the English-speaking Caribbean may be galvanized to collective action and use their disproportionate vulnerability to climate change to help shape the debate. While most of the states in this region may lack the institutional mechanisms to effectively address the threats posed by climate change, Mexico can credibly claim sufficient resources to do so. Cuba, meanwhile, has thus far demonstrated great institutional capacity to deal with near-term threats such as hurricanes but is otherwise subject to similar constraints as its neighboring states.

Cuba and Haiti

Cuba and Haiti offer good case studies for the region. Both are post-revolutionary, low- technology, and low-income societies. Cuba has a highly centralized sociopolitical structure, formed mostly around the military, while Haiti is highly decentralized. These two cases highlight the range of state capacity for much of the region.

Cuba has a well-structured system of research programs that covers a wide variety of problems focused on understanding the economic, technical, intellectual, and cultural development of the state. The country has the capacity to mobilize its population in advance of hurricanes that periodically strike the island. The Cuban military is the organizational core of the state, marshaling human and political capital through its networks. Its one million-strong *Milicia Territorial* is in charge of natural disasters and has operated with amazing efficiency, as is evident from the extremely low human toll during the most severe storms. In 2004, two hurricanes (Charley and Ivan) and a prolonged drought caused $3 billion in losses, 33 percent more than the earnings from tourism that year, and with virtually no loss of life.

Despite these strengths, Cuba lacks any significant capacity to recover from storm damage. Reasons for this include:

- Cuba has been unable to generate sufficient internal resources to invest for a sustained recovery.
- Cuba's international credit-worthiness is poor; there is limited access to external credits and loans.
- It is impossible to restrict consumption further to divert resources to investment as consumption is already depressed.
- The enterprise management reform process (*perfeccionamiento empresarial*) is new and very slow in implementation and therefore the central control of enterprises and government functions remains.
- Political and social structural impacts hamper reconstructive efforts such as the decline in real wages, increase in disguised unemployment, steeper income stratification, and resource and wealth hoarded by the elites.

Worrying demographic trends further hamper Cuba's ability to address recovery and development issues. Cuba's population is declining due to low fertility, out-migration, and a dramatically aging population. Migratory trends have drastically altered the ethnic makeup of Cuba—70 percent of the Cuban population is black, mostly of Jamaican origin, up from 30 percent 60 years ago, and migration into the cities has resulted in extreme over crowding. Havana, which represents 0.67 percent of the nation's land mass, comprises 25 percent of the nation's population.

Haiti, on the other hand, is already a failed state, and faces an ecological crisis. There is virtually no state presence outside of Port-au-Prince, where 90 percent of all government employees live and work, leaving the rest of the country in the hands of non-governmental organizations. The Haitian government is among the weakest in the world in terms of providing essential public services.

Both the Haitian and Cuban cases contain lessons for states attempting to contend with the deleterious effects of climate change:

- Excessive state control and centralization of decisionmaking such as exists in Cuba might well work to prepare the populace in the face of natural disasters, but proves much less efficient in helping that state recover from the damages caused by those disasters.
- Complete decentralization and minimal state involvement, as in Haiti, neither prepares the populace nor has the capacity to recover after the storms.

Regional Implications

Should the societies in question fail to properly adapt to climate change, increased migration and competition for scarce resources may raise tensions between neighboring states to the point of conflict. Civil conflicts are more likely than state-on-state violence, though some conflicts may straddle this divide. Of the potential areas of future conflict exacerbated by the effects of climate change, three merit special attention:

The Dominican Republic and Haiti. There are multiple signs that the future trajectory of this relationship is unsustainable with the potential for an escalation in violence. Vastly different levels of development (with per capita GDP estimated at $8,100 versus $1,300 in 2008 in purchasing power parity terms), explosive population growth, unfettered out-migration from Haiti to the Dominican Republic, and deepening racial and cultural tensions all raise the risk of violent conflict. This could take the form of civil unrest in Haiti or actions taken against Haitians by the Dominican military. The fact that Hispaniola sits in a major storm path also raises the possibility of hurricane-induced catastrophe affecting one or both nations.

The Guatemala-Honduras-Nicaragua Zone of Instability. Although not as dire as the situation on Hispaniola, the Guatemala-Honduras-Nicaragua axis poses another regional flashpoint for many of the same reasons. The three states will likely experience rapid population growth against the backdrop of comparatively low levels of development (respectively $5,200, $4,400, and $2,900 in 2008 per capita GDP, figures which contrast sharply with Mexico at $14,200, and are lower than Belize at $8,600, and El Salvador at $6,200). Given the low adaptive capacity generally assessed for these states, and the projected population growth, extensive out-migration from Belize may combine with pressures on Mexico's southern border to provoke conflict. The relative exposure of Honduras and Nicaragua to powerful Caribbean storms further heightens the potential for disruption.

The US-Mexican Border. **Increases in the frequency and scale of natural disasters caused from climate change could have a threat multiplier effect on immigration to the United States. US immigration is principally rooted in issues of economic deprivation and disparity.** If certain areas of Mexico, Central America, and the Caribbean become uninhabitable—either due to rising sea levels and temperatures, or because traditional agricultural and water resources cannot be sustained—then pressure on the US-Mexican border will increase. The potential for new climate-related epidemics may also affect the prerogatives for border security. The US-Mexican border may emerge as a future flashpoint, not as an area directly affected by climate change, but rather as a force to contend with the intensified migratory patterns that result.

Overall Foreign Policy Implications

The region's relative impotence in shaping the future trajectory of global climate change, coupled with the likelihood that it will bear the brunt of some of the most severe repercussions, will likely exacerbate several troubling tendencies in the region's politics. In Central America and the Caribbean, these include populist measures to control domestic natural resources, greater suspicion and skepticism towards the United States, Europe, and traditionally dominant powers, and greater orientation towards powerful southern neighbors such as Brazil and Venezuela. Insofar as the predicted rapid growth of China and India contribute to high levels of carbon emissions, this may lead to a diminution of these states' relative authority and popularity.

If Mexico ceases to be a net oil exporter and does not harness the economic potential of new technologies, then it may falter as a regional force as well. This may result in Mexico, Central America, and the Caribbean becoming increasingly inward looking, consumed by internal problems, and remaining at the margins of global action on climate change.

In relations with the United States they will remain vigilant. The United States is the single greatest determinant of change in the region—from water division issues with Mexico to political hegemony and aid support. If the United States seeks collaborative and effective regional mechanisms to jointly manage the challenges of climate change, this may provide a new opportunity for strengthening relations.

The implications for US foreign policy are great:

- Providing ever-increasing humanitarian assistance to neighbors in need.
- Avoiding catastrophic waves of refugees.
- Avoiding the need to intervene militarily as so often done in the Caribbean.
- Gathering goodwill in this area of scientific expertise, goodwill that should be advantageous in serving other US geopolitical and national security purposes.

Cuba represents a potential US partner once Fidel Castro is no longer in power. Cuban doctors have lived and worked throughout the region, creating a goodwill network along with bolstering the local medical systems. Cuba's medical and pharmaceutical resources in the region are second only to the United States, and already well in place. The United States could partner with Cuba in the creation of a rapid deployment humanitarian force, using Cuban resources and US funds. The first step would be direct talks between the US and Cuban militaries, as the Cuban military is the most organized force in the state and has access to US forces through proximity at the Guantanamo Naval Base.

Additionally, the United States should partner with Mexico to contain carbon emissions and directly address climate change challenges. Having established relationships and plans in place will be essential in dealing with whatever is to come.

Climate Change Negotiations

In approaching climate change negotiations with Mexico, Central American, and the Caribbean states, the United States needs to bear in mind that in many cases governments in the region may lack the capacity or will to deliver on their policy promises. Moreover, with the exception of Mexico, most of the states in question are not significant greenhouse gas emitters. Belize is even a net remover of greenhouse gasses. Although the states of the region may readily sign on to a climate change agreement, their ability to affect global climate change is limited.

Mexico, along with India, China, and Brazil, feels that it represents the interests of the developing world. Mexico has the ability to take a global leadership position on climate change, more so than with its regional neighbors. It may sign a climate change agreement but likely will be unable to enforce it effectively.

End Notes

[1] This section is extracted from the Executive Summary of the Phase I report (see Scope Note): ***Mexico, Central America, and the Caribbean: Impact of Climate Change to 2030, A Commissioned Research Report*** (NIC 2009- 11), of November 2009. Some of the judgments in this report (Phase II) may differ from the Phase I report.

Chapter Sources

The following chapters have been previously published:

Chapter 1 – This is an edited, reformatted and augmented version a National Intelligence Council Special Report NIC 2009-11D, dated December 2009.

Chapter 2 – This is an edited, reformatted and augmented version a National Intelligence Council Conference Report CR 2010-03, dated January 2010.

INDEX

A

B

C

D

E

F

G

N

O

P

Q

R

S

T